Part Decomposition of 3D Surfaces

OTHER BOOKS BY DAVID L. PAGE, PH.D.

More Republic, Less Cowbell: Renewing the American Experiment through Institutions, Not Noise

The Art of the Compromise: Returning American Democracy to Better Days

Knox County 2012 Charter Review Committee: Personal Notes and Position Papers

Scruffy Little Essays: A Collection on Knox County 2007–08

Part Decomposition
of 3D Surfaces

A revised and reissued edition of the 2003 doctoral dissertation

David L. Page, Ph.D.

Warped Minds Press
Knoxville, Tennessee

Part Decomposition of 3D Surfaces

Originally submitted as a doctoral dissertation in 2003.
First trade edition published by Warped Minds Press in 2026.

ISBN: 979-8-9956290-0-9

Warped Minds Press
Knoxville, Tennessee

For the readers who still believe old work can speak anew.

Contents

List of Tables

List of Figures

Preface

I wrote the pages that follow as a doctoral dissertation in 2003 at the University of Tennessee, capturing a specific moment in computer vision and a specific moment in my own life. A dissertation serves a committee, while a book must serve a reader, and this edition carries the original work from one setting to the other. I chose not to rewrite the text. The document that follows remains, in structure and substance, the same work I defended in 2003. That decision reflects intent. I wanted recovery rather than revision. Time changes tools, habits, and entire workflows, yet certain questions continue to matter and certain ideas continue to hold. That belief brought me back to this work.

The dissertation did not wait in a clean archive. The files lived on aging media—CDs, obsolete file systems, and formats that modern tools no longer recognize without effort. One such artifact appears below.

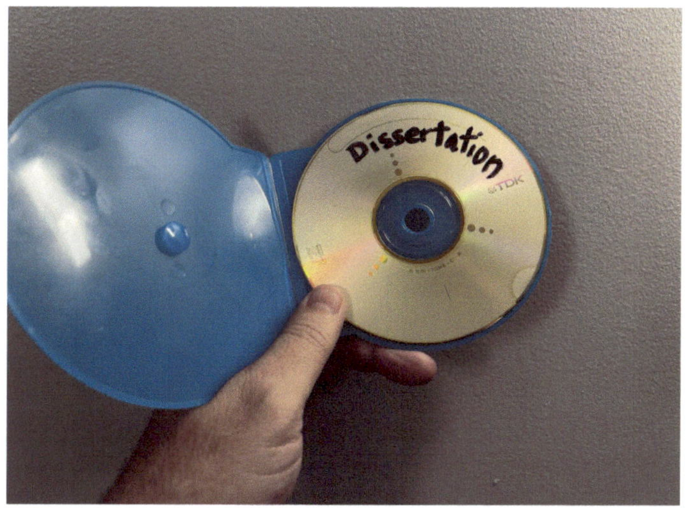

The original dissertation archive, preserved on CD.

That disc carried more than files; it carried a workflow. I began the recovery with modest expectations. A title page compiled, and a few sections appeared before the failures arrived. Figures disappeared, tables broke across margins, and hundreds of images surfaced only as EPS files—many produced with tools such as Xfig that no longer fit easily into modern environments. Each obstacle forced a translation between generations of software.

The work demanded patience. Old tools resurfaced, new tools filled gaps, and formats yielded one by one until the fragments aligned and the document returned. The effort revealed more than a manuscript. The effort revealed its construction—each tool, each assumption, and each constraint embedded in the work.

A dissertation records scholarship at a point in time, while a published book invites readers into that record. That distinction guided the preparation of the front matter and the presentation of the text that follows. The work remains what I completed in 2003. The difference lies in access.

David L. Page, Ph.D.
Knoxville, Tennessee

Acknowledgments

"IF I HAVE SEEN FURTHER IT IS BY STANDING ON THE SHOULDERS OF GIANTS" —SIR ISAAC NEWTON, 1676

Although I do not pretend to have seen as far as Newton—if I have even seen far at all, I do believe his quote is very apt in that many giants, not necessarily of the scientific variety, have afforded me the opportunity to pursue a graduate degree. To each of them, I am forever grateful.

First and foremost, I am deeply indebted to my family, especially my mother, Shirley Page, and my brother, Dan Page. Where I am today is in no small part due to their love and support. I also owe a special thanks to my Uncle Momo, Lon Boyd, who has always encouraged me to "get that education."

I additionally would like to thank my advisor, Dr. Mongi Abidi. His willingness to support my work and his guidance throughout my studies has allowed me to develop my skills as a researcher within a supportive team environment. I thank him for that opportunity. Also, I would like to thank, Dr. Paul Crilly. His advice and counsel over the years have been of equal importance. To Dr. Andreas Koschan, I say thank you as well for the many technical—and sometimes not-so-technical—discussions with regard to this research and life in general. I would further like to thank the other members of my committee: Dr. Daniel B. Koch, Dr. Conrad Plaut, and Dr. Hairong Qi. I greatly appreciate their time and input to this dissertation.

Within the IRIS Lab, I owe many thanks to both students and staff. Specially, I thank Vicki Courtney-Smith for her jovial smile and helpful nature that have always greeted any problem or deadline. Throughout my work, Justin Acuff and Tak Motoyama have routinely worked miracles, and for that they have my heartfelt admiration. To my fellow graduate students, I express my sincerest gratitude for the many conversations that have had a tremendous impact on my research and myself as a person. Faysal, Sun, Yan, Brad, Bernard, and Michael thank you so much.

Finally, I must express my appreciation to the many friends outside of my studies who have helped to relieve the sometimes stressful solitude of graduate school. In particular, Clint and Andrew have earned a special place in Heaven for putting up with me as a roommate. Your friendships through both good times and hard times have been a source of strength. To Molly and Marci, I can not say thank you enough for the many wonderful adventures as we followed the Vols from Pasadena to Timbuktu. I am profoundly grateful for your friendships as well. I also wish to thank the Ultimate Frisbee folks whose comaraderie each Wednesday evening has been invaluable. Last but not least, I would like to thank Lisa whose encouragement in the final days of this work have truly inspired me and have opened my eyes to the world beyond graduate school.

Chapter 1

Introduction

Let us begin with a brief exercise. Take a few minutes and glance around the room. Note objects in the room, perhaps items on your desk. Can you distinguish specific objects? Can you visually separate them from their background? For instance can you isolate the coffee mug on your desk from the clutter of books and papers? The answer is—of course—yes, you can and you do so with ease.

This seemingly simple task—deceptively simple—actually requires the coordination of millions of receptors within your eyes and billions of neurons in your brain. Photons of light bounce throughout the room and into your eyes striking each retina. Then rods and cones in the retina translate this light into neural signals and transmit them along the optic nerve. The optic nerve splashes this avalanche of input across neurons at the back of the brain—the visual cortex. These neurons fire sending ripples, like stones tossed into a pond, out to other areas of the brain and energize millions of neural networks. More neurons now fire igniting their own networks and again sparking new ripples. The mind orchestrates this rippling activity into a coherent thought that elevates to the conscious plane. The little voice in your mind replies, "There's my mug!" The graceful elegance of this complex process is truly a marvelous wonder—a wonder that allows you to isolate and identify the mug on your desk. Yes, all that for a mug.

Now, imagine that you are a *computer vision* engineer. How can you get a computer to do the same thing—separate the mug from the clutter of books and papers? How can you get a computer to decompose, or segment, a complex scene into simpler parts? In a very focused context, an answer to this question is the research goal of this dissertation. In particular, we have developed a novel *part decomposition*, or *part segmentation*, algorithm for surfaces. For input, instead of neural signals from a human eye, we will use triangle meshes generated from laser range scanners. For processing, instead of the neural networks of your mind, we will use segmentation algorithms implemented in a computer. Hopefully, our proposed algorithms mimic your visual perception, at least to a certain extent.

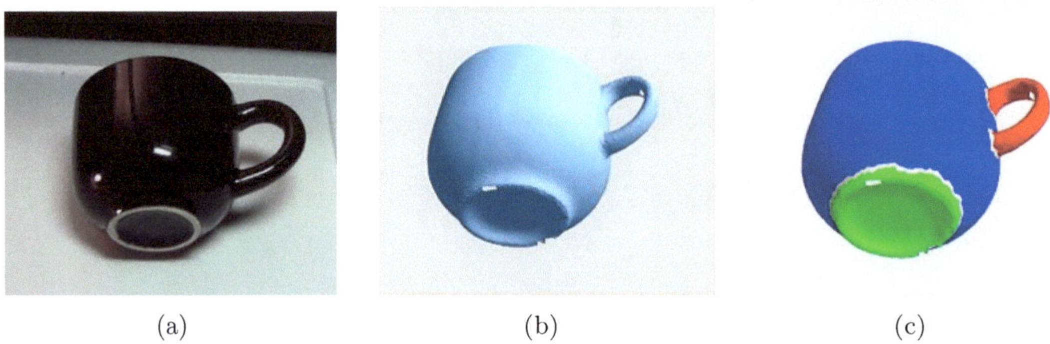

(a)   (b)   (c)

Figure 1.1: Part decomposition example for a coffee mug. (a) Photograph of the original mug. (b) A surface mesh model of the mug reconstructed from range images. (c) Part decomposition of the mug into three parts: a cup, a handle, and a base.

As a simple example of part decomposition, again think of the mug on your desk. Suppose the mug is similar to the one in Fig. 1.1(a) and that we can somehow generate a computer model of the mug as shown in Fig. 1.1(b). If your perception is like most viewers, then you decompose the mug into three or four different smaller parts. Most viewers would agree that the mug consists of a bowl-shaped cup, a handle protruding from the cup, and a base at the cup bottom. The color labels in Fig. 1.1(c) illustrate this segmentation. You have mentally decomposed the mug into three simpler parts. By extension, when we view more complex scenes such as the clutter on a desk top, we also decompose the scene into simpler parts. This example illustrates the objective of our research. In fact, our algorithm generated the segmentation shown.

This dissertation presents the details of our part decomposition algorithm. The remainder of this first chapter outlines the applications for our algorithm in Sec. 1.1 and the motivation for this research in Sec. 1.2. In Sec. 1.3, we then present the minima rule, which is the human vision theory that serves as the foundation of our research. To implement the minima rule, we briefly review the state of the art in Sec. 1.4. Next, we emphasize the contributions of this dissertation in Sec. 1.5, and we conclude with a block diagram of our system and the document organization in Sec. 1.6.

1.1 Applications

Segmentation, whose roots date back to the dawn of digital image processing, is an age-old problem in computer vision. As Marr (Marr, 1982) has stated, the goal of segmentation is to partition a data set into groups that are more meaningful. The difficulty is that this partitioning is not well posed and the term "meaningful" is highly subjective. Consequently, useful solutions are often *ad hoc* in origin. Marr points out that segmentation is a vague all-encompassing notion that typically digresses into a philosophical debate. In the context of image processing, he further argues that most images are too complex and often do not contain enough information for segmentation to succeed. Despite Marr's objections, segmentation has become a fundamental and ubiquitous topic in computer vision and in image processing, specifically (Gonzalez and Woods, 1993). The lesson to be learned from Marr however is that useful segmentation requires a precise formulation—or better yet, a formulation with strong philosophical support—of the segmentation goals. Such a formulation requires specific knowledge of the application at hand. To this end, we identify our application domain in the following paragraphs and in so doing assert our motivation for research. Additionally, as we will see, we are not interested in the traditional *image segmentation* problem but in the more general *surface segmentation* problem where we approximate a surface with a triangle mesh representation.

What is our application and why does it require segmentation? Our two applications are *scene modeling* and *reverse engineering*.

Scene modeling is the process of constructing a 3D computer model of a real-world scene where such models are useful in flight or driving simulators, architectural walk-throughs, and other virtual reality applications (Burdea and Coiffet, 1994). Research examples include urban landscapes (Frueh and Zakhor, 2001; Frueh and Zakhor, 2002), in-door environments (Yu et al., 2001), architectural structures (Faber and Fisher, 2002), industrial facilities (Johnson et al., 1995; Hebert et al., 1995), and precious art statues (Bernardini et al., 1999; Levoy et al., 2000). To illustrate, imagine a military simulator where a tank commander is training for a mission in an urban environment such as Mogadishu, Somalia. We could populate this simulation with cartoon-like models of buildings and roadways designed by a computer artist. To achieve convincing realism, an artist would methodically build-up the simulation from basic shapes such as boxes and cylinders. On the other hand, we could use the scene modeling techniques in (Frueh and Zakhor, 2001; Frueh and Zakhor, 2002) to rapidly model the streets of a specific city by driving through that city. Instead of an artist recreating a city-scape, we reconstruct it using a scene modeling system mounted on the roof of a van. As another example, imagine an art student in Knoxville who wishes to study the chisel patterns on Michelangelo's statues in Italy. Although she could travel to Europe, statue modeling such as (Levoy et al., 2000) offers a much more convenient alternative. She could simply download a 3D model of Michelangelo's David and use a virtual reality viewer to study the sculpture without ever leaving Knoxville. Creating computer models of buildings, rooms, and statues is the goal of scene modeling.

With reverse engineering instead of visually pleasing models, the objective is accurate as-built models of existing objects. Although reverse engineering is actually a broad field that encompasses many concepts, our specific definition is the ability to create a computer-aided design (CAD) model of a real-world part (Bernardini et al., 1999; Motavalli et al., 1998). By contrast, forward engineering is to create a real-world part from a CAD model. The automation of forward engineering, or computer-aided manufacturing (CAM), has significantly impacted recent technologies in system design. CAM has also introduced rapid prototyping into the design loop and facilitated changes on demand after the deployment of a design (Yan and Gu, 1996). The automation of reverse engineering, or computer-aided reverse engineering (CARE), promises to impact the design process in a similar fashion. CARE allows electronic dissemination of as-built parts for comparison of original designs with manufactured results. Additionally, CARE allows construction of CAD models of existing parts when such models no longer exist as when parts are out of production (Thompson et al., 1999). A military example of the potential for CARE is the Mobile Parts Hospital initiative within the U. S. Army Tank-automotive and Armament Command. The vision for the parts hospital is an emergency manufacturing unit for frontline deployment. Although the hospital should ideally have access to a CAD database, CAD models for a part may not necessarily be available such as for vehicles that have undocumented field modifications. A CARE scanner, however, allows even an untrained—in terms of engineering practices—soldier to create high quality CAD models. Additionally, a CARE scanner is a valuable tool for documenting part failures and thus creating an electronic history of the life cycle for a part.

1.2   Motivation

Although scene modeling and reverse engineering may seem dissimilar, they in fact share the common thread of *surface reconstruction*. Methods of surface reconstruction include (Hoppe et al., 1992; Hoppe et al., 1994; Edelsbrunner and Mücke, 1994; Delingette, 1994a; Curless and Levoy, 1996; Whitaker, 1996; Curless, 1997; Pulli et al., 1997; Amenta et al., 1998; Mencl and Müller, 1998; Bernardini et al., 1999; Gopi et al., 2000). Surface reconstruction is a two step process where by we first acquire the geometry of a scene or an object and then reconstruct its topology. The geometry acquisition is a digitization process whereby a sensor such as a coordinate measuring machine, a touch probe, a stereo pair, or perhaps a range scanner measures the location of points on the surfaces in a scene or on an object. Then, topology reconstruction finds the interconnection of these points. We refer to the collection of points as a *point cloud* and their interconnection as

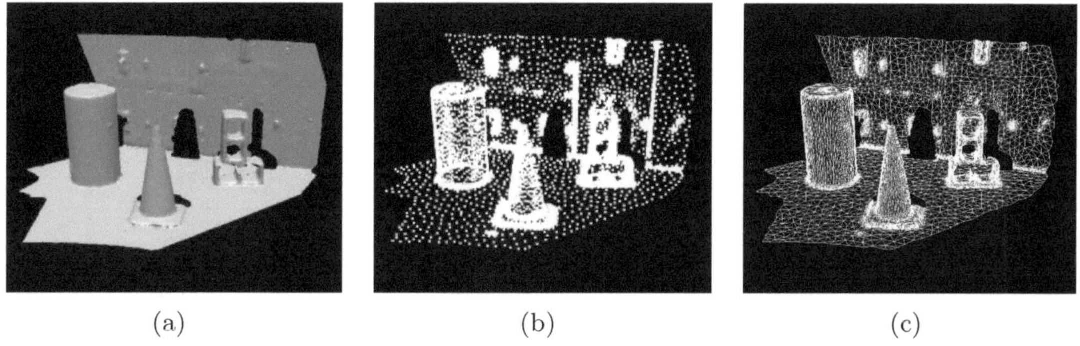

(a)   (b)   (c)

Figure 1.2: Scene modeling of an industrial scene. (a) The original scene with a barrel, cone, and blocks. (b) A point cloud derived from measurements of the scene. (c) A mesh reconstructed from the point cloud data. Notice that the mesh models the entire scene as a single connected surface—a blanket model.

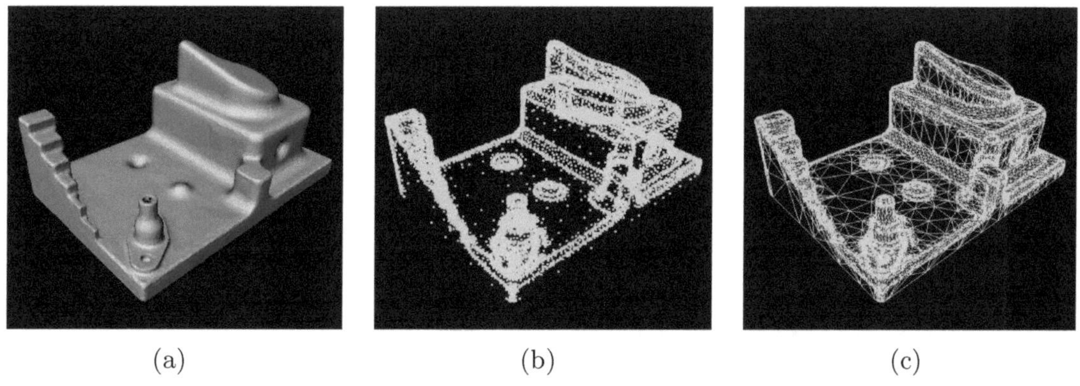

(a)   (b)   (c)

Figure 1.3: Reverse engineering example of a manufactured component. (a) Rendering of reconstructed part. (b) A point cloud derived from measurements of the part. (c) Underlying triangular mesh showing the blanket model.

a *surface mesh*, or simply a mesh. Consider Figs. 1.2 and 1.3 that show examples of the process.

Notice that the meshes in Figs. 1.2(c) and 1.3(c) are single contiguous surfaces. The meshes represent each connected object, that is to say objects that are physically touching each other, as one ubiquitous surface. By way of analogy, we describe this representation as a *blanket* model where Fig. 1.4 shows a simple illustration. Recalling our visual exercise, grab a blanket from your bedroom and lay it over your desk. The blanket will take the form of the desk, the books, the papers, and the mug.* Now, suppose we can apply an epoxy to the blanket so that

---

*In the case of a mug, our blanket analogy does breakdown, somewhat. Consider that the blanket can not change genus, without tearing it, to conform to the topology of the mug handle.

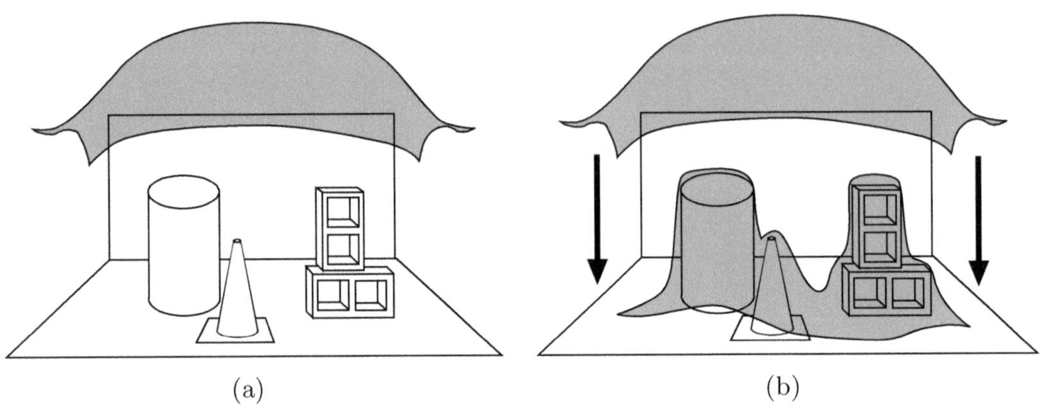

(a)   (b)

Figure 1.4: Reconstruction of a scene from multiple range images leads to a blanket model. These illustrations depict the analogy.

it hardens and thereby creates our blanket model. We can pick up the stiffened blanket and carry it with us. We can show it to other people. The problem is that if someone is only interested in the mug we have to give them the whole blanket. We really do not know which section of the blanket might contain the mug. Would it not be better if we could segment the blanket into smaller blankets—ones that are more manageable and more meaningful? We need a *mesh segmentation* algorithm. We need to decompose the blanket into smaller meaningful parts.

1.3 The Minima Rule

The term *meaningful* has cropped up again, echoing back to Marr. Although we know our applications are scene modeling and reverse engineering, we still need to identify what type of segmentation we expect. We need to counter Marr's objections and identify a theory to govern our segmentation. For scene modeling, we expect a segmentation that benefits real-time visualization of the scenes. We need to chop up the blanket so that the pieces are amenable to visualization. For reverse engineering, we expect a segmentation that leads to a more compact description of the object and that possibly facilitates the most grandiose of all computer vision tasks—the illusive task of object recognition. We look to the world of cognitive psychology for help. Researchers in human perception have identified a theory, known as the *minima rule* (Hoffman and Richards, 1984), that provides a precise formulation for segmentation. This rule defines meaningful in

---

So, we must tear the blanket and stitch it appropriately to truly model the mug. This point may seem minor but a reconstruction algorithm that accurately recovers topology is crucial and is an active area of research.

terms of human visual perception. If we segment our blanket mesh using a theory of human perception, the subsequent submeshes should naturally meet our needs for both applications. This point will become clear later. The marriage of the minima rule to scene modeling and reverse engineering is the primary impetus for our research. Is this marriage out in left field, or is it an important research pursuit? In the next section, we look at the state of the art in mesh segmentation to address this concern and to highlight the hole that our research fills.

1.4   State of the Art

Although image segmentation is a well known and thoroughly researched topic in computer vision, mesh segmentation has only recently become of interest where Mangan and Whitaker (Mangan and Whitaker, 1999) are perhaps the first to coin the term itself. In this section, we identify the current research on this topic and highlight a few shortcomings.

From a review of the literature, we have identified five papers that represent the state of the art. The first three are Vincent and Soille (Vincent and Soille, 1991), Wu and Levine (Wu and Levine, 1997) and Mangan and Whitaker (Mangan and Whitaker, 1999), whose methods directly address the mesh segmentation problem. The other two are Tang and Medioni (Tang and Medioni, 1999) and Taubin (Taubin, 1995), whose methods address curvature estimation. As we will see, curvature estimation is an essential component of the mesh segmentation algorithm that we propose.

Wu and Levine (Wu and Levine, 1997) are perhaps the first to directly attack the mesh segmentation problem as we have posed it. Their method uses electrical charge distribution equations to simulate the charge density on a mesh, and they identify segmentation boundaries as regions with the lowest charge density. This physics-based approach may seem unusual but their results are quite nice good for certain data sets. The strength of their algorithm is its robustness to measurement noise. The drawbacks however are that the method does not scale well to large data sets and that simulated charge distribution has limitations as a definition for segmentation. Wu and Levine suggest that their approach follows the minima rule—just as we propose to do—but in practice charge distribution does not directly relate to the rule. Additionally, the search algorithm they have implemented has certain limitations as well. Their algorithm tends to become trapped in local minima.

Mangan and Whitaker (Mangan and Whitaker, 1999) offer a different approach. The major contribution of their work is that they implement the well-known watershed algorithm from image processing on a mesh data structure. They reformulate the watershed algorithm from morphological image operations to gradient-following mesh operations. The capabilities of the watershed algorithm is a tremendous strength of their algorithm. The basis for their segmentation is the local curvature on the mesh. In particular, contours of high curvature bound areas of low curvature. They argue with heuristics that these boundaries offer a meaningful segmentation, but Marr's warnings about meaningful come to mind. From our literature review, we argue that high curvature boundaries indeed do not form a meaningful segmentation—at least for our applications—and we suggest that this approach is one drawback to their implementation.

Further, we suggest two other drawbacks. The first is the curvature estimation that governs their segmentation. The estimate they use is not robust to noise and in most cases leads to significant over segmentation. Also, since their method estimates Gaussian curvature, it is not useful for the minima rule, which requires estimation of the principal curvatures. The second drawback is their implementation of the watershed algorithm. Although Mangan and Whitaker demonstrate nice results, they have implemented a "bobsledding" version of watersheds where one initiates a segmentation with random seed points and follows the gradients from the seeds to watershed basins. This formulation is susceptible to local plateaus and thus requires post processing to handle over segmentation. As a result, Mangan and Whitaker implement an *ad hoc* solution to account for over segmentation, based on the depth of each watershed region.

Although Vincent and Soille (Vincent and Soille, 1991) propose a segmentation algorithm for 2D images, they generalize their algorithm to the arbitrary connectivity of a graph, such as a mesh. This algorithm does not appear in the review of Mangan and Whitaker, but it does propose a fast implementation of watersheds that is important to consider since the image processing literature devotes significant attention to this algorithm. Their graph variation is a flooding approach to mesh segmentation and as such has implementation advantages over Mangan and Whitaker. The downside to their algorithm is that it requires a pre-sorting of the mesh vertices according to water heights. As we will see, this pre-sorting is not suitable for our application of the minima rule.

Finally, since we are interested in the minima rule, curvature is important to our proposal as well. Unfortunately, as noted above, both curvature segmentation methods above (Wu and Levine, 1997; Mangan and Whitaker, 1999) have drawbacks with regard to their curvature estimations. Subsequently, we look to the literature for better methods. Tang and Medioni (Tang and Medioni, 1999) and Taubin (Taubin, 1995) represent the state of the art. Tang and Medioni offer a robust algorithm that estimates the sign of Gaussian curvature and the principal directions for noisy point clouds while Taubin presents an algorithm that estimates both principal directions and principal curvatures for triangle meshes. The drawback of Tang and Medioni is that they do not estimate principal curvatures while the drawback for Taubin is that he does not handle surface noise.

1.5   Contributions

The algorithms that we have developed extend the above state of the art. Wu and Levine present a robust curvature estimation method with a simple segmentation algorithm while Mangan and Whitaker present a robust segmentation algorithm with a simple curvature estimation method. We have developed an algorithm that has both traits—a robust curvature estimation method and a robust segmentation algorithm—and we ground our algorithm in the theory of the minima rule.

In particular, we have developed a **Minima Rule Decomposition Algorithm**, or more simply the **Minima Rule Algorithm**, that overcomes many of the drawbacks with Wu and Levine and Mangan and Whitaker. (A block diagram of this algorithm appears in Fig. 3.6.) The heart of this segmentation is two new algorithms known as **Normal Vector Voting** and **Fast Marching Watersheds**. Normal Vector Voting is a curvature estimation algorithm, and the Fast Marching Watersheds is a new implementation of the watershed algorithm for surface meshes. Finally, we have developed a **Part Saliency Metric** that handles any over-segmentation problems that may arise. To emphasize, this dissertation yields four contributions to the state of the art as follows.

Part Decomposition: Minima Rule Algorithm   The most significant contribution is the development of a computer vision algorithm that follows the human vision theory of the minima rule. To date, no computer vision algorithm implements the minima rule for 3D surfaces. As we have noted, Wu and Levine (Wu and Levine, 1997) do attempt an implementation, but their approach is not true to the minima rule theory since they do not use a proper curvature estimate. The algorithm that we present in this dissertation represents the first computer vision implementation of the minima rule for mesh segmentation.

Curvature Estimation: Normal Vector Voting   The second major contribution is the development of a robust curvature estimation algorithm known as Normal Vector Voting (Page et al., 2001; Page et al., 2003e). Although Tang and Medioni (Tang and Medioni, 1999) and Taubin (Taubin, 1995) offer important contributions, we have developed an algorithm that bridges the gap between these two algorithms. Our algorithm robustly estimates both principal directions and principal curvatures at the vertices of a triangle mesh, despite measurement error in creating the mesh.

Mesh Segmentation: Fast Marching Watersheds   The third contribution is the development of a mesh segmentation algorithm inspired by the popular watershed algorithm for image segmentation. We call our algorithm Fast Marching Watersheds. Although Mangan and Whitaker (Mangan and Whitaker, 1999) have demonstrated the feasibility of adapting image processing watersheds to surface meshes, their algorithm is a "bobsledding" approach that leads to significant over segmentation and requires handling of certain special cases. Similar to Vincent and Soille (Vincent and Soille, 1991), Fast Marching Watersheds avoids these problems by employing a "hill climbing" approach. Unlike Vincent and Soille, our algorithm does not require the pre-sorting step and thus does not require random access to each of the vertices in a triangle mesh. This difference is important to our application of the minima rule since we make local decisions about "water heights" and not global ones, as we explain in later sections.

Shape Measure: Part Saliency Metric   The final contribution is a new Part Saliency Metric, derived from a human vision theory (Hoffman and Singh, 1997). After we decompose a scene or object into a set of parts, we create a Part Adjacency Graph to define the relative relationship of each part. Our proposed metric assigns a value to the visual salience, or importance, of each part and to the salience of connections between parts. This metric enables filtering of oversegmentations that might occur where we merge the least visually salient parts with other more salient ones.

1.6   Document Organization

The remainder of this dissertation documents the details of our algorithms and the above contributions. Chapter 2 presents a survey of the literature for each contribution and also justifies our choice of the minima rule. Then, we overview the complete algorithm for part decomposition in Chapter 3. Next, Chapter 4 documents the theory that supports our Normal Vector Voting algorithm for curvature estimation. For mesh segmentation, we present our Fast Marching Watersheds algorithm in Chapter 5. To handle over-segmentation issues, Chapter 6 proposes a pattern vector algorithm to compute the salience of a part. After combining the theory from each of these chapters, we develop our part decomposition algorithm. The results from this integration are in Chapter 7. These experimental results demonstrate the robust capabilities of our algorithms and their successful application to a wide variety of objects and scenes. Finally, we conclude in Chapter 8.

Chapter 2

Literature Review

This chapter presents a review of the research literature. We begin with an investigation into part decomposition to establish our choice of the minima rule in Sec. 2.1. This review focuses on theories of human vision and computer vision in an effort to address both philosophical and implementation issues. With the minima rule, curvature estimation becomes important, and Sec. 2.2 discusses this topic. In Sec. 2.3, we review mesh segmentation algorithms to identify specific computer vision implementations that may be appropriate to the minima rule. Then, Sec. 2.4 reviews a variety of shape measures in order to quantify the quality of a segmentation. We conclude with a summary of the key articles in Sec. 2.5.

2.1 Part Decomposition

What do we mean by decomposition? When we view the mug in Fig. 1.1, how do we decompose it into simpler parts? How might a computer do so? Fig. 2.1 is one possible result. The question is what governs our decision process to decompose the mug in this manner. Does a theory of human vision exist that explains our choice?

2.1.1 Gestalt Grouping

At the turn of the century, gestalt* psychologists (Koffka, 1935; Wertheimer, 1958) began to formulate the idea that our minds group scenes. Wertheimer (Wertheimer, 1958) formalized a set of principles known as the gestalt principles of organization,

---

*A direct translation in English of the German word *gestalt* is often not adequate but usually this word translates as *form*. A better translation is *organized structure*. Gestalt psychologists emphasized perception in their experiments and observed the organized groupings that perception often yields (Kanizsa, 1979).

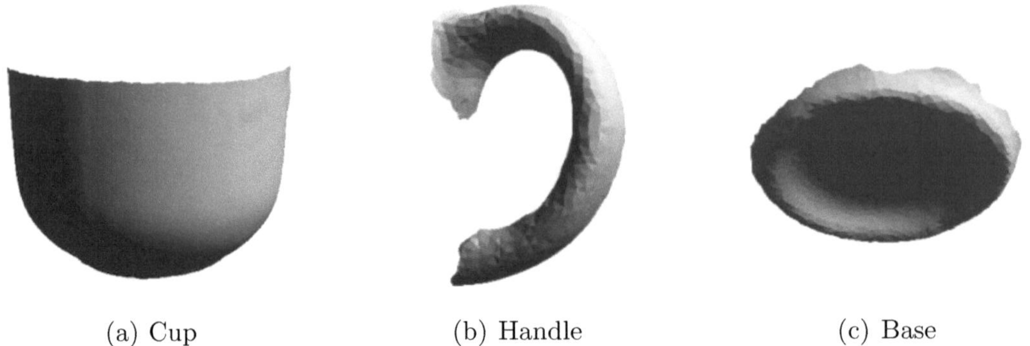

(a) Cup   (b) Handle   (c) Base

Figure 2.1: Visual parts that most observers see for the mug in Fig. 1.1.

which suggest that our mind and our perception tend to group our visual input. See Fig. 2.2. Along this line of thought, Palmer (Palmer, 1977) studied observer's abilities to recognize parts of figures. He demonstrated with simple line drawings that human perception tends to identify parts that follow the gestalt principles as in Fig. 2.3. These gestalt observations initiated interest into the notion that human vision groups—or decomposes—scenes and objects into simpler parts.

As time has passed, the gestalt school has faded—for a variety of reasons—and lost favor among cognitive psychologists. The gestalt insights, however, have focused attention on the importance of organizational groupings of scenes in human perception. Although the gestalt principles emphasized simple lines and dots to highlight such groupings, more recent cognitive research (Marr, 1982; Marr and Nishihara, 1978; Hoffman and Richards, 1984; Hoffman and Singh, 1997; Biederman, 1987; Tversky and Hemenway, 1984; Juttner et al., 1996; Rosch et al., 1976; Koenderink and van Doorn, 1982; Palmer, 1977) has emerged that addresses complex groupings in images. This research, which parallels new developments in digital imaging, has led to a growing consensus that decomposition of shapes into their constituent parts is fundamental to human vision and—by extension—to computer vision.

Although it may seem obvious, we do see the world in terms of parts and the early stages of our perception function primarily to identify these parts. We term this visual process *part-based decomposition*, or more simply *part decomposition*. For a fair assessment, we do note that some researchers such as Cave and Kosslyn (Cave and Kosslyn, 1993) do argue against the notion of parts, but these objections are not predominant. Naturally, many researchers (Pentland, 1989; Pentland, 1987; Pentland, 1986a; Biederman, 1987; Biederman, 1985; Guzman, 1971; Binford, 1971; Terzopoulos et al., 1987; Brooks, 1981; Dickinson et al., 1992;

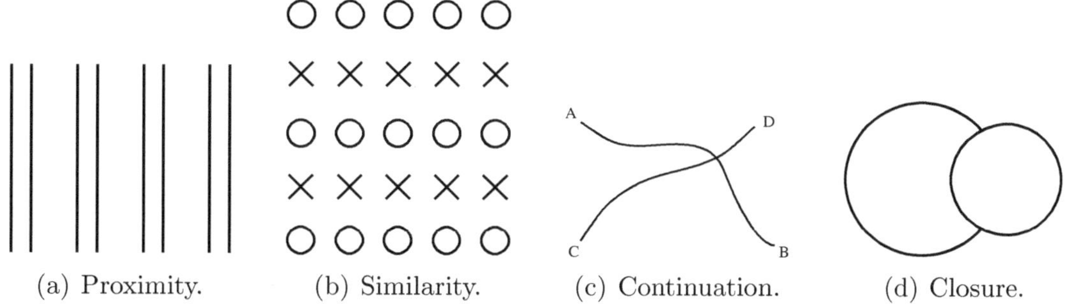

(a) Proximity.   (b) Similarity.   (c) Continuation.   (d) Closure.

Figure 2.2: Gestalt principles of organization (Andreson, 1995). (a) With the *principle of proximity*, we perceive four pairs of lines rather than eight separate lines, individually. Our mind tends to group items that are closer together. (b) With the *principle of similarity*, we tend to see two rows of X's and three rows of O's. Why? Why not five columns of alternating X's and O's? Our mind seems to group elements that are alike into common units, in this case rows. (c) With the *principle of continuation*, we perceive two lines from A to B and from C to D. It seems unnatural for our mind to perceive the lines as A to D and C to B although this grouping is just as valid. (d) With the *principle of closure*, we observe one circle that occludes another. In this figure, only one complete circle exists; the other is incomplete. Yet, we tend to believe that one circle is sitting on top of the other one and that we indeed see two circles.

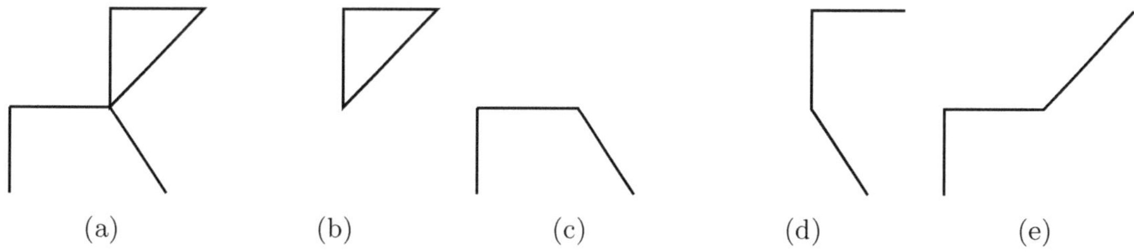

(a)   (b)   (c)   (d)   (e)

Figure 2.3: Line drawings used by Palmer (Palmer, 1977) for studying part decomposition. (a) Original example object. (b,c) Part decomposition of (a) that follow the gestalt principles. (d,e) Part decomposition that does not follow these principles (Andreson, 1995). Note that (b,c) seem more natural while (d,e) somehow seem unnatural.

Siddiqi and Kimia, 1995) in the computer vision community[†] also argue that parts are essential to computer vision tasks as well. These arguments have motivated us to explore part decomposition as the starting point for our research.

2.1.2 Definition of a Part

Now that we are convinced that parts are important, what do we mean by a *part*? Is a part a functional component such as an arm or a leg? Is it a geometrical entity such as a box or a cylinder? Perhaps it is a perceptual feature such as a handle that protrudes from a mug. These questions lead us to classify part decomposition algorithms based on their notion of a part (Vaina and Zlateva, 1990; Wu and Levine, 1997; Hoffman and Singh, 1997). In particular, as (Hoffman and Richards, 1984; Vaina and Zlateva, 1990) suggest, we categorize these algorithms as either *primitives-based* or *boundary-based* methods. As an aside, both (Hoffman and Richards, 1984) and (Vaina and Zlateva, 1990) actually use the term primitive-based instead of primitives-based. We distinguish these terms since primitive-based seems to imply a simple or less cultured approach when the actual intent is that these methods employ a library of basic primitives. Also, we note that Vaina and Zlateva offer a third classification, called axis-based. These algorithms such as (Blum, 1973) and (Blum and Nagel, 1978) rely on axes of symmetry and are thus only appropriate for objects that exhibit strong symmetry. For this reason, we do not consider these algorithms. Fig. 2.4 illustrates our categorization for part definitions.

2.1.3 Primitives-Based Methods

Primitives-based methods decompose an object or a scene first by defining a set of basic shapes—or primitives—and then by finding these shapes in the data. The predefined primitives are volumetric models that fully specify the 3D shape of each part. They are not points or contours. They are volumes. A primitives-based algorithm forms an initial configuration using a set of primitives that closely models the input data. Through an iterative search, the algorithm then scales, rotates, translates, removes, adds, and possibly deforms the configuration and each primitive in it until reaching a stopping criteria. The final set of primitives are the parts of the scene or object of interest.

[†]Throughout this dissertation, we have used the terms *human vision* and *computer vision* without strictly defining them. We do so now to avoid confusion. Human vision refers to the cognitive processes of the human mind and the inter-relationship between our eyes and our brain that allow us to see, model, describe, and recognize the world around us. Computer vision, on the other hand, is our attempt—often a meager one—to simulate human vision using computer algorithms.

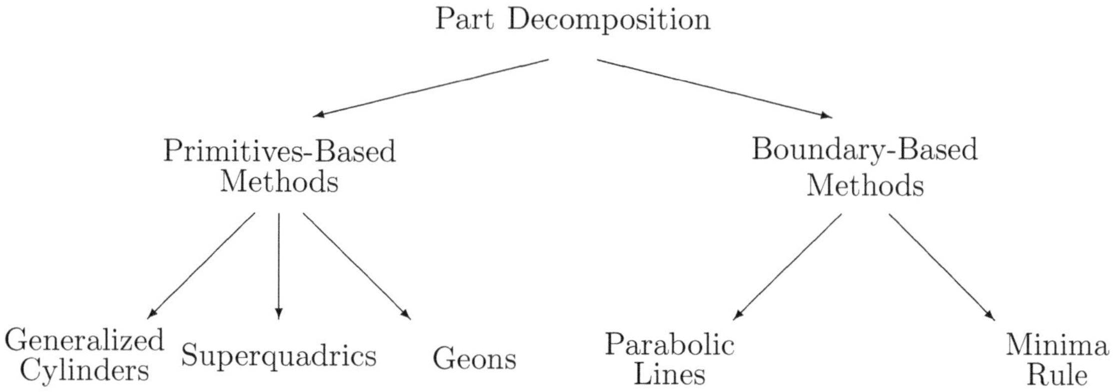

Figure 2.4: Categories of part decomposition algorithms.

The construction of the primitives dictionary is crucial to the success of these algorithms where a diverse set of shapes is important. The research literature offers an extensive set of possible primitives, but most are derivatives to one degree or another of the following:

- Generalized cylinders (Binford, 1971; Brooks, 1981; Marr and Nishihara, 1978; Nevatia and Binford, 1977),

- Superquadrics (Barr, 1981; Pentland, 1986b; Pentland, 1987; Pentland, 1989; Bajcsy and Solina, 1987), and

- Geons (Biederman, 1985; Biederman, 1987).

Generalized cylinders are historically the first methods proposed and thus have greatly influenced the field. Other methods have followed with most recent work focusing on superquadrics. Both generalized cylinders and superquadrics are parametric models that are defined by a strict set of mathematical equations. A decomposition algorithm iteratively adjusts the parameters of these equation to fit the primitives to the input data. For the generalized cylinders, the parameters define a closed planar contour and an axis. They form a volume by sweeping the planar contour along the axis thereby creating a generalized cylinder. To change the shape of the cylinder, one simply redefines the contour or warps the the sweep axis. Unfortunately, the number of variables to specify uniquely the contour and the axis can be quite large. As a result most practical implementations restrict the parameters to a more focused subclass.

To avoid the parameter overload problem, another solution is the superquadric model. These primitives provide as much flexibility as generalized cylinders but with fewer parameters. Superquadrics have greater mathematical support formalized through Barr (Barr, 1981), and thus are a more elegant solution. Pentland (Pentland, 1986b) introduced this family of primitives to the computer vision community with significant work following his research.

Geons are similar in spirit to both generalized cylinders and superquadrics but are a more qualitative approach. The previous approaches have well-defined quantitative parameters that govern the shape of the primitives. Geons use qualitative language, as opposed to a mathematical formulation, to describe their shape. Biederman (Biederman, 1985; Biederman, 1987) proposed geons—geometrical icons—as a set of primitives that are defined in terms of invariant image features. Biederman notes that certain properties of visual features remain invariant to perspective transformation through small angles. For example, a straight line in 3D appears straight in a 2D image and a curved line appears curved. Only by an accident of view does a curved line appear straight. Biederman presents four qualitative invariant features that result in a database of 36 geon primitives. These features are edge type, symmetry, sweep variation, and axis type, but he does not define these parameters using strict mathematical variables. Subsequently, most implementations of geons use generalized cylinders or superquadrics as the actual primitives in a computer algorithm. Geons are more of a theory of part decomposition than an implementation.

An interesting aspect of each of the above methods is that part *description* is inherent to the decomposition process. Not only do these methods yield a decomposition of a scene into parts but they also provide either a mathematical description, in the case of generalized cylinders and superquadrics, or a qualitative description, in the case of geons.

2.1.4   Boundary-Based Methods

As a departure from the above techniques, boundary-based approaches advocate that decomposition alone should precede description and not include it (Hoffman and Richards, 1984). Unlike primitives-based approaches, boundary-based methods attempt to decompose a scene or an object by identifying the boundaries between adjacent parts instead of matching primitives to the parts. As Hoffman and Richards (Hoffman and Richards, 1984) describe, a boundary is where we would draw a contour between parts, as if with a felt marker. A boundary is a contour on a surface where one part ends and another part begins. Decomposition involves finding these contours. Thus, we do not need to know what the parts look like, rather we only need to know where the parts intersect each other.

So what constitutes a part boundary? This question has led to significant debate within the human vision community, which has further led to different implementations in the computer vision community. Koenderink and van Doorn (Koenderink and van Doorn, 1982) kicked off this debate by proposing parabolic lines as boundaries. They argue, as evidenced in works of art, that humans perceive 3D shapes as composed mainly of elliptic regions with hyperbolic patches as glue between these regions. A basic teaching tool in art for drawing human figures is to have students sketch ellipses for faces, arms, legs, torsos, feet, and hands and then to stitch these ellipses together as the final drawing progresses. The intersection of elliptic regions and hyperbolic patches forms parabolic lines that Koenderink and van Dorn deem to be part boundaries. These lines occur where the patches transition from elliptic to hyperbolic—from positive to negative Gaussian curvature.

Hoffman and Richards (Hoffman and Richards, 1984) note a few problems with the above boundary definition. First, parabolic lines are invariant to figure and ground reversal. Figure and ground (Rubin, 1958) are a common way in cognitive psychology to distinguish the two sides of a curve in 2D and a surface in 3D. One side is figure; the other is ground. Consider the classic sketch of (Rubin, 1958) in Fig. 2.5(a). This sketch illustrates either two human faces that are nose to nose or just a single vase. The image that we see depends on how our mind chooses figure and ground. The reversal of which allows us to see the other image. Parabolic lines on a surface are invariant to figure and ground reversal where these lines are the loci of points with zero Gaussian curvature. Such points do not change whether we are on one side or the other of a curve or surface. When we view Fig. 2.5(a), however, we not only see different images with figure and ground reversal, but also we subsequently see different parts for each image. Hoffman and Richards voice another objection with regard to the unnatural parts that are sometimes spawned by parabolic line boundaries or—in some instances—no parts at all. They provide examples where experimental evidence shows that human observers choose boundaries that differ significantly from the parabolic line boundaries.

As an alternative Hoffman and Richards (Hoffman and Richards, 1984) propose their own definition of a boundary—the *minima rule*. The formal definition of this rule is as follows:

minima rule All negative minima of the principal curvatures (along their associated lines of curvature) form boundaries between parts (Hoffman and Singh, 1997).

As an illustration of this rule, refer to Fig. 2.5 again. In Fig. 2.5(b), we see how one choice of figure and ground and the minima rule lead to parts of a face. Similarly, in Fig. 2.5(c) we see how a different choice leads to parts for a vase. Experimental

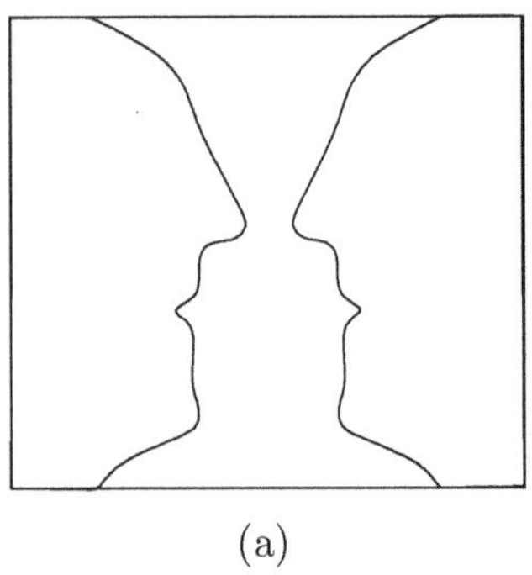 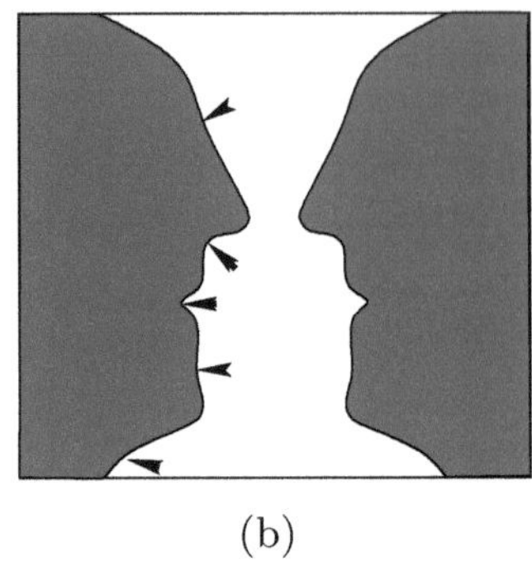 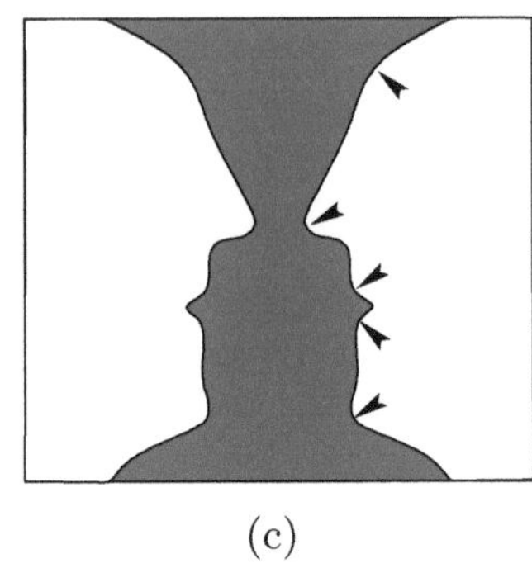

(a)   (b)   (c)

Figure 2.5: Visual perception sketches illustrating how figure and ground effect choices of parts. (a) What do you see? (b) If we choose the white space as ground, we see two faces that are facing nose to nose. (c) If we chose the dark space as figure, we now see a vase. In (b) and (c), the arrows denote the minima rule boundaries for parts (Hoffman and Richards, 1984; Rubin, 1958).

evidence (Baylis and Driver, 1995b; Baylis and Driver, 1995a; Braunstein et al., 1989; Driver and Baylis, 1995; Hoffman, 1983) further supports the rule.

Researchers have explored other definitions of part boundaries. For example, Hoffman and Singh (Hoffman and Singh, 1997) cite "deep concavities" (Marr and Nishihara, 1978), "sharp concavities" and "concave regions" (Biederman, 1987), and "limbs and necks" (Siddiqi and Kimia, 1995). Additionally, Fischler and Bolles (Fischler and Bolles, 1986) propose high-curvature points as part boundaries similar to Mangan and Whitaker's implementation (Mangan and Whitaker, 1999). The minima rule precisely captures and more accurately formalizes these definitions into a single concise rule, and it overcomes the limitations of Koenderink and van Doorn's parabolic lines.

Finally, the major strength of the minima rule with regard to our applications is that it is computable, robust, and invariant (Hoffman and Singh, 1997). In short, this means that we can implement the rule as a computer algorithm and that the rule is applicable to a wide variety of situations.

Computable implies that the theory has a mathematical foundation. As a counter example, geons are not computable since geon theory is a qualitative and not quantitative description of parts. For the minima rule, curvature is the computable measure that we need to implement. We note however that humans do not necessarily compute curvature directly, but rather they probably use visual cues such as shape from shading. Regardless, the minima rule is a computable theory of part boundaries and as such is important to practical implementation.

Another important trait of the minima rule is that it is *robust* to shape variability. Consider our mug example. Mugs come in many shapes and sizes but as a general rule mugs consist of a cup and a handle—two parts. The handles may vary from the elaborate ornamentations on a German stein to the simple curves of a household coffee mug. The minima rule precisely defines the boundary of the handle and cup for each of these examples and for many more as well. The algorithm does not require *a priori* information of the type of object or scene and is a general theory for any shape. This robustness is an important characteristic for our application domains.

The third strength is the *invariance* of the minima rule. Although this trait mainly applies to the 2D version of the rule for images, it has some significance to 3D surfaces. With respect to images, invariance means that the rule survives under perspective transformations. For 3D, invariance means that the rule is independent of scale and rotation changes. A big mug and a little mug both have a handle and a cup as parts. Scale does not change our perception of the part boundaries. An upright mug and an upside-down mug also have the same parts. Rotation does not change the part boundaries. The minima rule holds true for each of these situations.

For these reasons, we select the minima rule as the foundation for our decomposition algorithm. The implementation of this rule requires the development of a curvature estimation algorithm and a mesh segmentation algorithm. We review the literature for these topics in the next two sections.

2.2   Curvature Estimation

We have identified robust curvature estimation as a weakness in the current literature with regard to triangle meshes. In particular, we have not found a method that robustly estimates both principal curvatures and principal directions, which we need for implementing the minima rule. Our research intends to address this issue. Most research in the literature addresses curvature estimation in the context of range images with little work available for the more general problem of surface meshes. Since image processing and mesh processing require different tools, we do not intend to address the direct estimation of curvature from range images. Our interest instead is to address the more general problem of curvature for a surface mesh. We refer the interested reader to (Flynn and Jain, 1989) and (Suk and Bhandarkar, 1992) for excellent surveys into curvature-from-range methods.

2.2.1   Differential Geometry

As background, we first present a brief overview of surface curvature in the important context of differential geometry (do Carmo, 1976; O'Neill, 1997). The

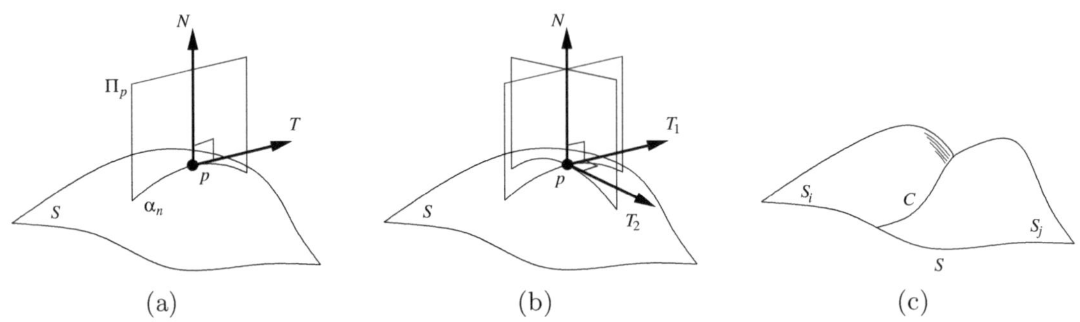

(a)   (b)   (c)

Figure 2.6: Illustration of curvature for a surface. (a) Shows a normal curvature on the surface S at the point p. The plane Π_p contains the unit surface normal N and the unit tangent vector T for point p. (b) The principal directions T_1 and T_2 form an orthonormal basis for the infinite set of normal curvatures at p. (c) Crease discontinuity. The smooth surfaces S_i and S_j meet at crease C. At the crease, the minimum principal curvature follows the crease and has some finite value. The maximum curvature is infinite and orthogonal to the crease.

curvature of a surface intrinsically describes the local shape of that surface. Consider Fig. 2.6. The point p lies on a smooth surface S, and we specify the orientation of S at p with the unit-length normal N. We define S as a manifold embedded in $\mathbf{R}^3$. We can now construct a plane Π_p that contains p and N, and the intersection of Π_p with S forms a contour α on S. For this contour, there is a unique arc parameterized by length s, $\alpha(s)$, where $\alpha(0) = p$ and $\alpha'(0) = T$. This parameterization has the property that T is the unit-length tangent vector at p. With this construction, we now have a parameterized contour on S, and thus we can find the curvature of that contour. We define the *normal curvature $\kappa_p(T)$* of S at p in the direction of T as $\alpha''(0) = \kappa_p(T)N$. The normal curvature is for a single contour on S passing through p. This curvature $\kappa_p(T)$ does not however specify the surface curvature of S at p.

For surface curvature, we need to do a little more work since Π_p is not a unique plane. If we rotate Π_p around N, we form a new contour on S with its own normal curvature. We can see that we actually have an infinite set of these normal curvatures around p in every direction. Fortunately, herein enters the elegance of surface curvature. For this infinite set, we can construct an orthonormal basis $\{T_1, T_2\}$ that completely specifies the set. The eigenvectors, along with their associated eigenvalues, of the second fundamental form for S at p are a natural choice for this basis. These eigenvectors $\{T_1, T_2\}$ are the *principal directions* of S at p and are the directions of the tangent curves with maximum curvature $\kappa_p^1 = \kappa_p(T_1)$ and minimum curvature $\kappa_p^2 = \kappa_p(T_2)$, which are the respective eigenvalues of the second fundamental form. These maximum and minimum curvatures are known as the *principal curvatures* and lead to the following relationship for any normal curvature at p:

$$\kappa_p(T_\theta) = \kappa_p^1 \cos^2(\theta) + \kappa_p^2 \sin^2(\theta) , \qquad (2.1)$$

where $T_\theta = \cos(\theta)T_1 + \sin(\theta)T_2$ and $-\pi \leq \theta < \pi$ is the angle to vector T_1 in the tangent plane. The principal directions along with the principal curvatures completely specify the surface curvature of S at p, which is what we are seeking.

Combinations of the principal curvatures lead to other common definitions of surface curvature. The first of these—and perhaps the most common—is *Gaussian curvature*, which is the product of the principal curvatures $K = \kappa_p^1 \kappa_p^2$. This definition highlights negative surface curvature that occurs at hyperbolic patches since these patches occur where only one principal curvature is negative. The second definition of curvature is *mean curvature*. We specify mean curvature as the average sum of both principal curvatures $H = \frac{1}{2}(\kappa_p^1 + \kappa_p^2)$. Although neither Gaussian nor mean curvature specify the orientation of curvature, they are common definitions found in the estimation literature, as we will see in the next section. We emphasize however that we are not interested in just the principal curvatures, which lead to the Gaussian and mean curvatures, but also the principal directions.

Our challenge in estimating curvature is unfortunately that we are not dealing with a completely smooth surface such as S above but rather a piecewise-smooth surface $\check{S}$ where we apply the definition of a piecewise smooth as in (Biermann et al., 2000). The surface $\check{S}$ may for example be the union of three smooth surfaces S_j, S_k, and S_l, manifolds embedded in $\mathbf{R}^3$, such that $\check{S} = S_j \cup S_k \cup S_l$. We assume that S_j, S_k, and S_l are *orientable manifold surfaces, possibly with piecewise-smooth boundaries* (Kinsey, 1993; Biermann et al., 2000), and that their subsequent union $\check{S}$ also conforms to this same definition of a surface. The subsequent piecewise-smooth surface has discontinuity contours $C_{jk} = S_j \cap S_k$ where two smooth surfaces join as in Fig. 2.6(c). Other discontinuities occur at corner points $c_{jkl} = S_j \cap S_k \cap S_l$ where three or more surfaces join. Both principal curvatures are singular for such corners.

Another challenge is that we are not actually working with $\check{S}$ but rather with the mesh M that approximates $\check{S}$. Recall Figs. 1.2(c) and 1.3(c). We specify M as the pair $M = (K, V)$ where K defines the topology and V defines the geometry. We assume as with $\check{S}$ that M is an *orientable triangulated manifold surface, possibly with boundary* (Kinsey, 1993). The vertices V are samples of $\check{S}$ such that noise may corrupt these samples. We have the following

$$v = p + e \qquad (2.2)$$

where $v \in V$ is a specific vertex of M, $p \in \check{S}$ is a sample point on $\check{S}$, and e is a noise vector that accounts for measurement, registration, and isosurface extraction error. We can think of V as a point cloud and K as the interconnection of V to form the edges and faces of the triangles in M.

Our review of the research literature reveals that only a few papers address the issue of curvature estimation on triangle mesh such as M. Of those papers we have identified three classes of techniques:

- Surface Fitting Methods (SFMs),

- Direct Curvature Methods (DCMs), and

- Curve Fitting Methods (CFMs).

We discuss each class briefly with special emphasis on the curve fitting methods as those offer the most promise.

2.2.2 Surface Fitting Methods

SFMs fit an analytic surface to the data of interest and then use differential geometry to compute curvature from that function. With some modifications, we can use many of the analytic methods for range images listed in (Flynn and Jain, 1989) and (Suk and Bhandarkar, 1992). An interesting modification of (Flynn and Jain, 1989) for meshes is the approach in (Sacchi et al., 1999). Sacchi et al. fit spheres to adjacent triangles and use the Gaussian curvature of these spheres as curvature estimates. For a vertex, they average the curvature estimate for adjacent triangle pairs around the vertex. Most extensions of the range methods, however, require a local parameterization of the surface similar to the parameterization that an image provides for range data. Although a surface may allow many different functional representations locally, no practical global parameterization is useful. The approaches of (Hagen et al., 1998) and (Rössl et al., 2000) provide possible choices for local parameterization. Also, (Rössl et al., 2000) use thresholds and morphological operations on a mesh to identify smoothness discontinuities after estimation of curvature. Once we have a local parameterization, we can use methods such as linear regression (Ferrie and Levine, 1988; Flynn and Jain, 1988; Sander and Zucker, 1986) or splines (Naik and Jain, 1988; Vemuri et al., 1986) to estimate curvature. A more recent paper (Pulla et al., 2002) uses a local fit of a biquadratic polynomial and applies smoothing to improve the analytic estimate of curvature. In a similar approach, Yang and Lee (Yang and Lee, 1999) locally fit parametric quadric surfaces.

Instead of triangle meshes, some reconstruction methods generate smooth surfaces directly. With smooth surfaces, we can directly apply differential geometry to compute curvature. Such reconstruction methods include polynomial surface (Sapidis and Besl, 1995), splines (Eck and Hoppe, 1996), and subdivision surfaces (Hoppe et al., 1994), as examples. Unfortunately, these methods form C^2 continuous patches with C^1 stitching between these patches. The location of these stitches are arbitrary and may not follow the piecewise smooth seams of the original surface. Thus, curvature along stitch junctions is not straightforward. Extraordinary points of subdivision surfaces also require special treatment for curvature estimation (Reif and Schröder, 2000).

2.2.3 Direct Curvature Methods

DCMs are another class of algorithms. These algorithms use the topology and geometry of the mesh *directly* to estimate curvature. Since a triangle mesh is a piecewise-flat surface, the direct computation of local curvature is seemingly paradoxical (Mortenson, 1997). The curvature is singular at each point on the surface—infinite at vertices and edges and zero on triangle faces. We can, however, refer to the total curvature for regions on these surfaces, which is not necessarily singular.

Lin and Perry (Lin and Perry, 1982) use the angle excess around each vertex to estimate the total Gaussian curvature. Angle excess itself is well known with (Mortenson, 1997) providing a nice discussion in the context of computer graphics and the Gauss-Bonnett Theorem. We find another application of angle excess in series of papers (Delingette, 1994a; Delingette, 1994b; Delingette, 1997; Delingette, 1999). He lays out a framework for a surface representation that he calls a simplex mesh that is a dual to a triangle mesh. He discusses the total mean and the total Gaussian curvature for this surface representation and shows these formulations are directly related to angle excess for a triangle mesh. We find another angle excess approach in the discrete minimal surface and straightest geodesic work of Polthier and his coauthors (Polthier and Schmies, 1998; Pinkall and Polthier, 1993). Following this line of research, Desbrun et al. (Desbrun et al., 1999) define a curvature normal vector as a discrete definition of mean curvature for triangle meshes. As with the angle excess methods, Desbrun et al. use interior angles of triangles for their formulation.

With a different approach, Gourley (Gourley, 1998) presents a total *pseudo* curvature based on the dispersion of face normals around a vertex while Mangan and Whitaker (Mangan and Whitaker, 1999) refine this measure as the norm of a covariance matrix for these face normals. This pseudo curvature is proportional to the magnitude of Gaussian curvature. A novel algorithm from Wu and

Levine (Wu and Levine, 1997) is a physics-based approach where they simulate the distribution of charge density across a mesh. They relate this charge distribution to surface curvature. This approach also yields a pseudo curvature measure that is monotonically increasing relative to Gaussian curvature.

2.2.4 Curve Fitting Methods

We finally consider the CFMs. With these methods, we fit a family of curves individually around a point and then use the ensemble to estimate curvature. Martin (Martin, 1998) proposes a method that selects vertex triples from a mesh and fits circles to those triples. Tookey and Ball (Tookey and Ball, 1997) describe a more sophisticated method that uses five points instead of three but is only valid for data on a regular grid. Várady and Hermann (Várady and Hermann, 1996) present an algorithm for computing principal curvature from a collection of surface curves using a linear system. A very interesting paper (Tang and Medioni, 1999) proposes a novel approach to infer the sign of Gaussian curvature and compute principal directions from noisy data. This method is an evolution of Medioni's tensor voting theory (Medioni et al., 2000), which uses circular curves to discern features from a point cloud. A recent improvement to their original paper is (Tang and Medioni, 2002). From the Duplin indicatrix, Chen and Schmitt (Chen and Schmitt, 1992) formulate a quadratic representation of curvature at each vertex and then derive the principal curvatures using a least squares minimization of the resulting overdetermined system. Inspired by this approach, Taubin (Taubin, 1995) developed an algorithm that defines a symmetric matrix that has the same eigenvectors as the principal directions and eigenvalues that are related by a fixed homogeneous linear transformation to the principal curvatures. He estimates this matrix in discrete form for a triangle mesh using vertex pairs that share a common edge. In the context of surface reconstruction, Gopi et al. (Gopi et al., 2000) extend Taubin's algorithm beyond adjacent vertex pairs to arbitrarily close pairs and use a different weighting scheme. Another improvement to Taubin is (Hameiri and Shimshoni, 2002). We finally note the curvature work in the context of mesh simplification in (Heckbert and Garland, 1999). This paper outlines the relationship of the quadric error metric (Garland and Heckbert, 1997; Lindstrom and Turk, 1998) for triangle normals to curvature.

The SFMs require the most computational effort since they typically employ optimization in the fitting process. This optimization does provide some robustness to noise but does not inherently deal with discontinuities. The DCMs on the other hand are more computationally efficient but are more susceptible to noise errors. The exception is the method of Wu and Levine (Wu and Levine, 1997) that does demonstrate robust results. None of the DCMs, however, directly estimate the principal directions or principal curvatures that we seek. The CFMs are the most promising of the three classes. In particular, Tang and Medioni (Tang and Medioni, 1999) and Taubin (Taubin, 1995) offer unique contributions. Tang and Medioni's method is robust but their algorithm does not estimate principal curvatures, only principal directions. They construct a matrix—similar to Taubin—whose eigenvectors relate to the principal directions but they do not show how the eigenvalues relate to the principal curvatures. As stated above Taubin's algorithm does. Taubin relates the eigenvalues to the principal curvatures. As we will show in the next chapter, our contribution is to extend both Tang and Medioni's and Taubin's methods with a new algorithm that employs a geodesic neighborhood, a voting scheme, and Taubin's discrete formulation to generate robust results. In the next section, we explore potential mesh segmentation algorithms to see how we can implement the minima rule.

2.3 Mesh Segmentation

In the first section of this chapter, we reviewed several decomposition theories where we have identified the minima rule as the primary theory for our applications. This choice has led us to investigate curvature estimation in the previous section. We now review the literature on *mesh segmentation* to see how we might implement the minima rule in a practical system. We seek an algorithm to segment a mesh M that approximates a manifold surface $\check{S}$, embedded in 3D. Recall the mesh segmentation of a mug in Figs. 1.1 and 2.1.

2.3.1 Convex Polyhedra

An active area of research that is similar to mesh segmentation is *convex decomposition* in computational geometry. This problem seeks to decompose a non-convex polyhedron into smaller convex ones. The motivation for this work is to improve computer graphics such as rendering and shading (Chazelle et al., 1997). The seminal paper in this field is Chazelle (Chazelle, 1984) and his follow up articles (Chazelle and Palios, 1990; Chazelle et al., 1997; Chazelle and Palios, 1997). Other researchers (Bajaj and Dey, 1992; Hershberger and Snoeyink, 1998; Tang et al., 2000) have also contributed with significant interest growing. Most of these algorithms seek to find the simplest decomposition possible but Lingas (Lingas, 1982) shows the minimum decomposition complexity is NP-hard. Thus heuristics are necessary, and as a result the main focus of the research attempts to bound the worst-case complexity of the problem. A computer vision-based approach to convex decomposition is (Svensson and Sanniti di Baja, 2001; Svensson and Sanniti di Baja, 2002). This method is less rigorous than the computational geometry approaches above but yields nice results. A volumetric distance transform guides the segmentation process.

Although this research fits nicely with the scene modeling goals for real-time visualization, it does not address the needs of reverse engineering. The description of an object as a collection of convex parts is not very meaningful. Additionally, these algorithms assume the non-convex polyhedra are ideal models with no measurement error corrupting them. These ideal models are common in most computer graphics applications but not computer vision ones. For practical scene modeling or reverse engineering, measurement error degrades the quality of the mesh, and at present, convex decomposition does not address the effects of this noise. The growing interest in this research, however, does emphasize the importance of mesh segmentation to computer graphics and thus serves as a context for our research.

2.3.2　Range Images

Another area related to the mesh segmentation problem is range image segmentation where an example appears in Fig. 2.7. Hoover et al. (Hoover et al., 1996) survey the traditional approaches and establish a framework for comparing these algorithms. Suk et al. (Suk and Bhandarkar, 1992) also provide a review and present the fundamental groundwork for the problem itself. Some important papers are (Fan et al., 1986; Hoffman and Jain, 1987; Besl and Jain, 1988). More recent work includes (Baccar et al., 1996; Burgiss et al., 1998; Yang and Lee, 1999; Alrashdan et al., 2000; Froimovich et al., 2002). Of these methods, the watershed segmentation of (Baccar et al., 1996) is noteworthy. Baccar et al. use image processing watersheds and data fusion techniques to identify surface discontinuities. As we discuss later, we have an interest in watershed segmentation as well. In terms of part decomposition, the algorithm of (Froimovich et al., 2002) demonstrates nice results. They add an additional level of complexity beyond segmentation of range images by assigning functionality to the resulting part decompositions.

The drawback of range image segmentation, however, is that images are not surfaces. With a range image, one can exploit the regular row-column structure of the image and thereby simplify algorithm development. A typical surface mesh, on the other hand, does not offer this regular structure and to impose such a structure is often not practical. A mesh M has arbitrary connectivity in K. Recall the parameterization discussions in the review of curvature estimation techniques.

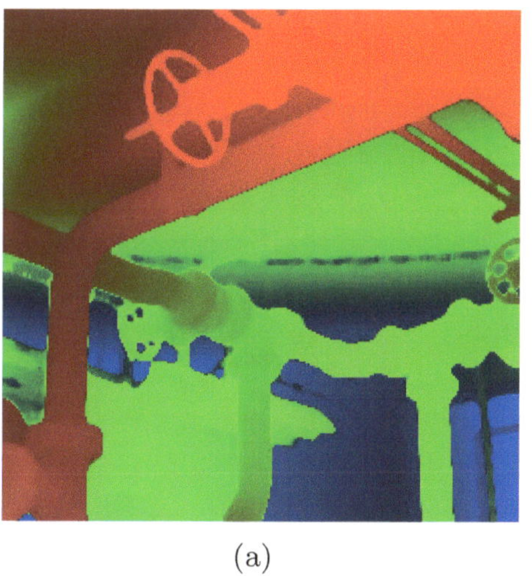

(a)   (b)

Figure 2.7: Example of range image segmentation for a scene of water pipes. (a) The initial range image with false coloring. Close surfaces are red while distant ones are blue. (b) A segmentation of the range image based on surface discontinuities.

2.3.3 Surface Meshes

So far, we have seen that convex decomposition from computer graphics and range image segmentation from computer vision do not address our application domain adequately. We truly need a mesh segmentation algorithm. Since this topic is relatively new, the literature only offers a few algorithms.

Two early mesh segmentation algorithms are (Falcidieno and Spagnuolo, 1992) and (Hebert et al., 1995). Falcidinieno and Spagnuolo segment a mesh into similar curvature regions of concave, convex, planar, and saddle patches. Hebert et al. compute quadric surface patches for curvature estimation and then employ a region growing method for segmentation. The region growing is a modification of (Faugeras and Hebert, 1986) that Hebert et al. have adapted for the arbitrary connectivity of a surface mesh. Johnson et al. (Johnson et al., 1995) demonstrate an application of this method. Another region growing method for CAD applications is (Sapidis and Besl, 1995). A recent computer vision system from Yu and Malik (Yu et al., 2001) presents a pipeline for reconstructing and editing scenes from range scans. The segment scene geometry into distinct surfaces to aid other tasks such as registration.

In computer graphics, Gregory et al. (Gregory et al., 1999) propose an interactive segmentation for morphing applications that requires user selection of feature points. Using the points as landmarks, they segment the mesh into morphing patches. Li et al. (Li et al., 2001) describe edge contraction and space sweeping to decompose a mesh for collision detection during visualization. Tan et al. (Tan et al., 1999) demonstrate decomposition results through a vertex-based simplification algorithm. Werghi and Xiao (Werghi and Xiao, 2002) propose a computer vision algorithm with applications in computer graphics to segment 3D scans of the human body. Their algorithm uses posture recognition as a first index into identifying human body parts. In a reverse engineering application, Rössl et al. (Rössl et al., 2000) define morphological operators such as opening and closing on the surface meshes and use these operators to segment surface discontinuities.

As noted previously, a very successful algorithm is Wu and Levine (Wu and Levine, 1997). Recall that they draw from the rich field of finite element analysis and implement segmentation as a physics-based approach by simulating electrical charge distributions over a surface mesh. Although this formulation yields a robust curvature estimate, the mesh segmentation algorithm they propose is somewhat simplistic. Their algorithm defines the triangle mesh as a direct connection graph. Then, they identify concave extremum within the graph and march from these nodes to neighboring nodes with lowest charge density. This march continues until they trace a closed contour across the surface of the mesh. The drawback to this algorithm is that it fails when an object has three or more parts that intersect at a common point. For such parts, a simple closed contour topology is not sufficient. Another drawback is that the algorithm relies on a local marching procedure that is susceptible to local minima. Under certain surface topologies, the algorithm becomes trapped and is unable to complete a closed contour.

A second algorithm that we consider in depth is (Mangan and Whitaker, 1999). Again, recall that they implement the watershed algorithm from image processing to a mesh data structure. As an improvement to their algorithm, we also note more recent papers (Pulla, 2001; Pulla et al., 2002). In image processing, the watershed algorithm is a well known thresholding method (Castleman, 1996). The basic idea is to view a data set as an elevation terrain where the height of the mountains and valleys corresponds to the maximum and minimum values of the data. Segmentation involves figuratively pouring water over this terrain and monitoring where the catchment basins of the terrain form. The algorithm subsequently groups areas of land with common basins, i.e. common watersheds, as segmentation regions.

The final algorithm that we need to consider is the image processing algorithm of Vincent and Soille (Vincent and Soille, 1991). This algorithm is also a watershed algorithm like Mangan and Whitaker but a different approach. In fact, the discussion in Vincent and Soille is an excellent review of watersheds for image processing. They conclude their paper with a section that extends their methods to graphs with arbitrary connectivity, including a brief discussion of 3D surfaces. Their algorithm uses two steps. The first one sorts each pixel in an image according to water height, and then the second step grows watershed regions

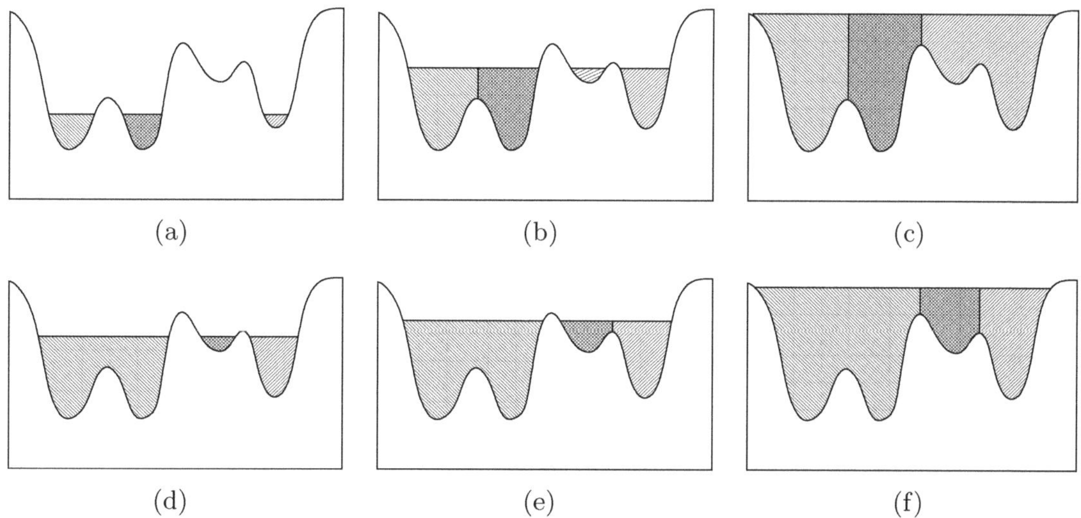

Figure 2.8: The flooding variant of the watershed algorithm is dependent on the initial choice for the marker set. (a) An initial marker set (b–c) Progression of the segmentation. Note the merge operation between (b) and (c). (d) A different choice for the marker set. (e–f) The effect of this choice on the subsequent segmentation.

by accessing this initial sorting and using a queue to enlarge from the bottom up. More recently, Rettmann et al. (Rettmann et al., 2000; Rettmann et al., 2002) implement this algorithm as a true mesh segmentation for cortical surface in medical applications.

As Vincent and Soille (Vincent and Soille, 1991) note, two different strategies are common for implementation. The first strategy is a bottom-up approach where we form catchment basins and flood the data with water. See Fig. 2.8. The second strategy is a top-down approach where we descend, or bobsled, down to the catchment basins from the slopes and ridges of the terrain. See Fig. 2.9. Both of these approaches require an initial threshold to create what is often called the marker set. This set is critical to the success of the algorithm since it defines the proper number of regions for the final segmentation. The initial boundaries of this set are too small to be a complete segmentation, but as the algorithm progresses these boundaries expand. For the flooding approach, when two boundaries come in contact as in the left side of Fig. 2.8(b), they are not allowed to merge. On the right side of that same figure, notice that a new basin has formed that has no correspondence, yet, to any other basin. As flooding continues in Fig. 2.8(c), we do merge this new basin with one of the original marker basins. These figures illustrate the importance of the marker set. As additional examples, Figs. 2.8(a) and 2.8(d) show two different marker sets with the corresponding final segmentations in Figs. 2.8(c) and 2.8(f). We see very different results.

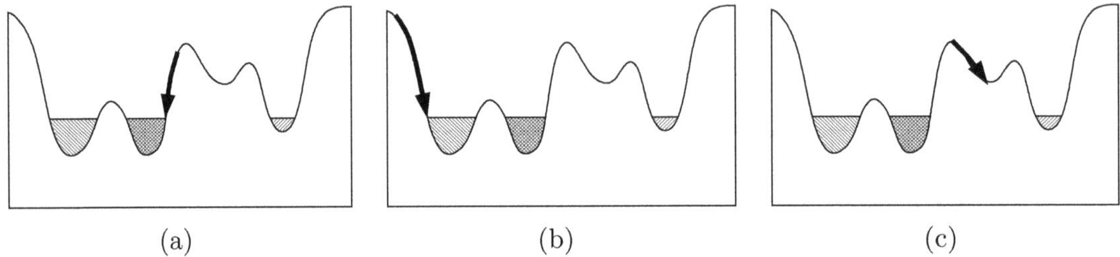

(a)   (b)   (c)

Figure 2.9: We show the marker set in each illustration. The dark arrows show the bobsled analogy where points on the data follow the gradient of the data to the catchment basin. (a–b) A typical bobsled. (c) A local minimum trap that requires merging to an appropriate catchment basin.

For the bobsled approach, the problem is a little different. As before we establish a marker set, but instead of flooding, we follow the gradients of the data to a marker set. Unfortunately, local minima may stop our bobsled from reaching any marker regions as in Fig. 2.9(c). Such cases require special consideration. Mangan and Whitaker (Mangan and Whitaker, 1999) have successfully implemented the bobsled method for their mesh segmentation algorithm. They note that the local minima problem leads to significant over segmentation and that such results are not very useful. To address over segmentation, they have developed an intricate filter-and-merge algorithm. This algorithm looks for watersheds with relatively low "water depths" and merges such regions with neighboring ones. Unfortunately, we suggest that this approach is an *ad hoc* solution and a more robust approach is necessary. Additionally, we believe that the bottom-up flooding approach to watersheds leads to better results. For this dissertation, we have developed such an approach, and to handle over segmentation we use a theory of part saliency as an alternative to the *ad hoc* water depth as in (Mangan and Whitaker, 1999; Rettmann et al., 2000). In the next section, we explore the concepts of part saliency.

2.4   Shape Measure

As stated in the last section, segmentation algorithms must handle the problem of over segmentation. If we look at the state of the art, Mangan and Whitaker (Mangan and Whitaker, 1999), for example, compute the depth of a watershed for each segmentation region and then merge regions that are relatively shallow. They assume that such regions represent areas of over segmentation. Other approaches (Wu and Levine, 1997) merge regions with relatively small surface area, based on the size of the segmentation region. These solutions are *ad hoc* and, as such, are not directly applicable to our desire to implement the minima rule.

With an implementation of the minima rule, proper segmentation—in other words segmentation without over segmentation—yields "good" parts. So, if we could measure the "goodness" of a part, we could simply filter bad parts when over segmentation occurs. The question that obviously arises is how do we measure, or even compute, part "goodness".

An answer to the question of "goodness" is *part salience* (Hoffman and Singh, 1997). Hoffman and Singh propose a human vision theory in the context of the minima rule that defines the salience of a visual part where salience is the visual significance of a part in terms of perception. They argue that parts help us index our cognitive memory of shapes during visual tasks. They further suggest that the saliency of a part determines its efficacy as an index. Efficacy relates to reaction times, error rates, confidence ratings, and decisions of figure and ground for our visual processing (Hoffman and Singh, 1997). Leveraging psychophysical experiments, Hoffman and Singh propose that the salience of a part depends on primarily three factors: relative size of the part, degree of protrusion, and strength of part boundary. Although they give quantitative definitions of these factors, their work is a human vision theory and not a computer vision algorithm. In particular, they only give a brief discussion of part salience for 3D surfaces and instead focus most of their attention to 2D silhouettes.

2.4.1 Shape Complexity

The challenge that our research addresses is the development of an algorithm that computes Hoffman and Singh's notions of part salience for 3D meshes. Since no algorithm in the literature implements Hoffman and Singh's definitions directly, we have broadened our literature review to include algorithms that measure shape. Hoffman and Singh's definitions are essentially measures to quantify the shape of a part. In the literature, we do find a diverse array of applications from satellite imagery (Oddo, 1992) to neuron morphology (Cesar and Costa, 1997) that require shape measures. To gain an understanding of these approaches and to inspire our search for a part salience measure, we review the literature in shape measures in the following paragraphs.

To begin, King and Rossignac (King and Rossignac, 1999) present a paper that is directly relevant to triangle mesh data sets. In their work, the authors consider lossy mesh compression and propose a shape measure to evaluate compression methods. Their shape factor is relative to a sphere of a given radius and measures the level of tessellation that a certain shape requires. Toussaint (Toussaint, 1991) proposes another measure for shape in 2D based on polygon decomposition. Toussaint argues that the number of interior triangles from the resulting decomposition that do not share an edge with the boundary of the polygon serves as a measure of shape. Chazelle and Incerpi (Chazelle and Incerpi, 1984) have proposed the sinuosity as a measure where sinuosity is the number of times that a polygon's boundary alternates between complete spirals of opposite orientations (Toussaint, 1991). Although polygon triangulation is a 2D problem, convex decomposition of polyhedra is the 3D analog with examples in (Chazelle, 1984; Hershberger and Snoeyink, 1998) as outlined in the previous section. A downside to these algorithms and the subsequent shape measures based on them is the computational effort required to compute the decompositions. A computationally efficient approach to measuring shape is the polyhedra moments found in (Li, 1993). Li presents efficient methods for computing various degrees of moments for a polyhedron. A more recent paper is (Osada et al., 2002), which proposes a method of computing shape signatures of polygonal models. The key to their algorithm is the definition of an appropriate shape function based on a global geometric property. They suggest the distance between two random points on a surface is one possible function.

Spatial database systems (Bryson and Mobolurin, 2000) as related to geographic information systems are another area of research where quantifying shape has gained interest. A theoretical characterization of polygonal objects common to spatial databases is through the fractal dimension. A notable investigation is Mandelbrot's paper (Mandelbrot, 1967) that applies fractal analysis to Britain's shoreline. As a shift away from fractal dimensions, Brinkhoff (Brinkhoff et al., 1995) proposes a pattern vector approach to measuring shape. Brinkhoff develops a metric based on a set of descriptive parameters such as notches, vibrations, and convexity. He combines these parameters in a weighted sum to compute the complexity of the object. Bryson and Mobolurin (Bryson and Mobolurin, 2000) further expand on these concepts.

We also find shape measures in biological shape analysis, particulary cell morphology. An interesting paper by Cesar and Costa (Cesar and Costa, 1997) develops a multiscale approach. In the context of neuron morphology, they formulate a normalized multiscale bending energy description of neural cell boundaries. Bending energy, itself, is a measure of shape where Young et al. (Young et al., 1974) use bending energy to characterize the contours of biological objects. Vliet and Verbeek (van Vliet and Verbeeck, 1993) extend bending energy definitions to 3D data sets.

From the computer vision and image processing literature, a few key methods use information theory (Shannon, 1948) to measure shape. Oddo (Oddo, 1992) has developed a segmentation algorithm based on global shape entropy to extract building boundaries from aerial imagery. The entropy definitions in Oddo follow from the gray level definitions of entropy in (Pal and Pal, 1989). Oddo uses a region growing technique to identify building shapes where the curvature of the region boundary defines the entropy function. Roui-Abidi (Roui-Abidi, 1995) also uses curvature and entropy but in a different context. She formulates a curvature-entropy measure relative to Oddo and uses her formulation to govern sensor placement to maximize information in sensor views. To estimate the uncertainty in pose for multiple sensor views, Stoddart et al. (Stoddart et al., 1998) propose a registration index as a means of quantifying the error that one might expect when registering a particular shape. This registration index represents some level of shape measure. Finally, Gahli et al. (Ghali et al., 1998) define a metric for the amount of rotational information that an image contains and investigate rotational information properties of Latin character sets. A more recent computer vision method is the linear shape descriptor in (Sanniti di Baja and Svensson, 2002), which seeks a skeletal description of shape volumes. Another recent paper is (Athitsos and Sclaroff, 2002) that describes a method for computing protrusion of fingers in images for hand shape classification.

2.4.2 Part Salience

Unfortunately, none of these methods are independently sufficient for a straightforward development of Hoffman and Singh's part saliency into a computer vision algorithm. The pattern vector formulation of Brinkhoff (Brinkhoff et al., 1995), however, does offer an approach to how we might combine the three factors that Hoffman and Singh outline while the computational techniques of (Li, 1993) and (Athitsos and Sclaroff, 2002) suggest possible methods of computing these factors. As we will see in a later chapter, we have developed a part saliency metric that follows the theory of Hoffman and Singh that fills this gap in the literature.

The fundamental principle of Hoffman and Singh is the relative significance of adjacent parts on an object or in a scene. Their research identifies measurable quantities relative to the parts that determines the overall salience of a part. We illustrate these measures in Fig. 2.10 for simple 2D sketches. These figures demonstrate that global properties such as the size of each part in Fig. 2.10(a) and the degree of protrusion in Fig. 2.10(b) effect the salience. In general, a large part is more salient that a smaller one. A part that protrudes significantly is more salient that a part that does not. Further, these figures illustrate local properties that effect salience such as the turning angle at a part boundary in Fig. 2.10(c). Hoffman and Singh extend these ideas and present mathematical definitions for both 2D contours and 3D surfaces. We propose to use these salience measures as a more appropriate filter-and-merge method than the *ad hoc* approaches in Mangan and Whitaker and Wu and Levine.

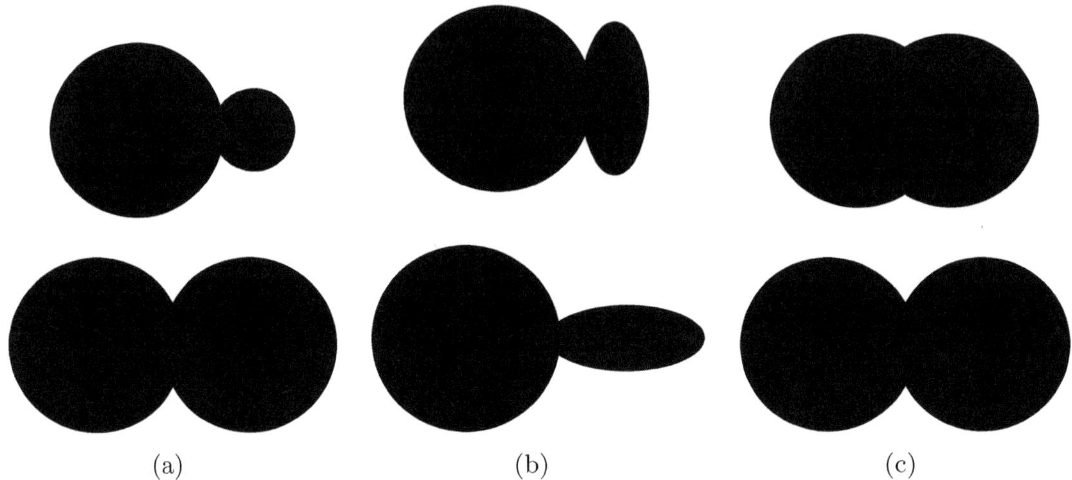

(a)   (b)   (c)

Figure 2.10: These illustrations are simple examples that demonstrate part salience. Each one shows two parts connected with a different salient feature. The upper objects have two parts that are less salient than the parts of the lower objects. (a) Relative size of parts determines salience. (b) Degree of protrusion. (c) Turning angle of the cusp boundary.

2.5 Summary

This chapter has presented an extensive review of the literature for each of the topics explored in this dissertation. We now highlight the key articles that serve as the foundation for the theories and algorithms that we develop in the next chapters. Below we delineate these articles for clarity and note how we intend to extend the state of the art.

Part Decomposition After surveying a variety of human vision theories, we have selected the minima rule (Hoffman and Richards, 1984) as the most appropriate approach for scene modeling and reverse engineering applications. Our contribution is to develop a computer vision algorithm—the Minima Rule Algorithm—that adheres to this theory.

Curvature Estimation With the minima rule, curvature estimation becomes an important topic. We have identified tensor voting (Tang and Medioni, 2002) and Taubin's algorithm (Taubin, 1995) as the most promising methods in the literature. We bridge the gap between these two algorithms to extend the state of the art.

Mesh Segmentation To segment a mesh using the minima rule and a curvature estimation algorithm, we have found three algorithms (Vincent and Soille, 1991; Wu and Levine, 1997; Mangan and Whitaker, 1999) that serve as starting points for our development of a new mesh segmentation algorithm. In particular, our contribution is to develop a watershed segmentation algorithm for triangle meshes that implements the flooding model of watersheds.

Shape Measure A successful segmentation algorithm requires appropriate methods to handle over segmentation. We have explored the literature with regard to shape measures in an effort to measure the segmentation quality. Although we have not find an adequate computer vision algorithm that satisfies our needs, we have identified part saliency (Hoffman and Singh, 1997), which is a human vision theory, as a potential solution. Our contribution is again to develop a computer vision algorithm for this theory, as with the minima rule.

This summary list completes the literature review. In the next chapter, we develop our Minima Rule Algorithm that extends the above state of the art.

Chapter 3

Part Decomposition: Minima Rule Algorithm

The primary contribution of this dissertation is the development of the Minima Rule Algorithm for decomposition of triangle mesh approximations of 3D objects and scenes. In this chapter, we briefly overview this algorithm. As the previous chapter outlines, a significant body of research is available from cognitive psychology, in particular human perception, that seeks to understand the fundamental elements of human vision with regard to part decomposition. From our survey of the literature, we have selected the minima rule theory from Hoffman and Richards (Hoffman and Richards, 1984) as the most applicable to our problem domains for scene modeling and reverse engineering. This chapter is a short presentation of the Minima Rule Algorithm as a computer vision system. We note that we do not discuss in detail the elements of this algorithm, here. We delay such specifics to subsequent chapters that focus on individual components of the system. Our goal for this chapter is to introduce the algorithm and build the foundation for the remaining chapters.

We begin this chapter by introducing in Sec. 3.1 a simple example the we use extensively throughout the remaining chapters. This example is illustrative of the concepts that we develop. In the next section, Sec. 3.2, we present the input and output for the Minima Rule Algorithm. Finally, we conclude in Sec. 3.3 with a block diagram of the complete system. This diagram shows the key components that we investigate in detail during later chapters.

3.1 Mug Example

Throughout our discussions, we use the same example of a coffee mug to illustrate the various concepts of our algorithms. Photographs of the actual mug we

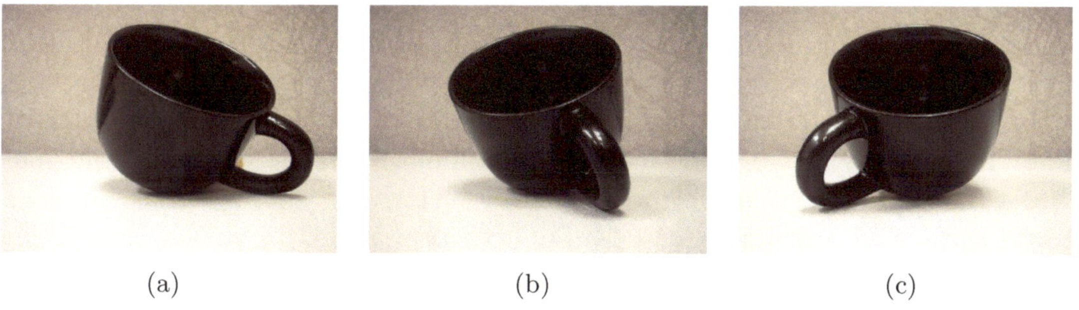

Figure 3.1: This sequence of photographs shows the actual mug used as an example throughout this dissertation.

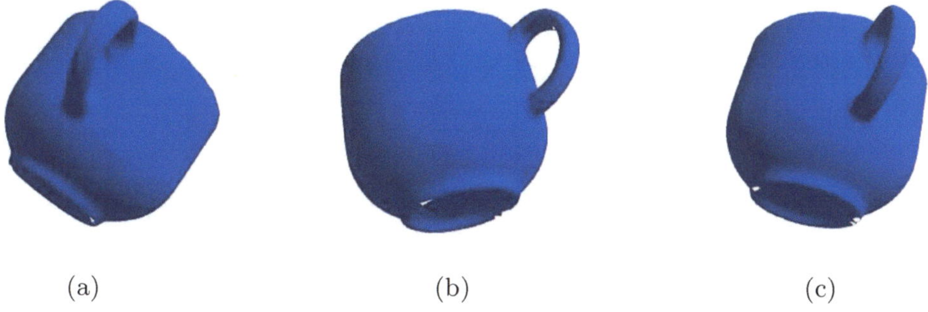

Figure 3.2: This sequence of images show computer renderings of the triangle mesh that models the mug.

have chosen appear in Fig. 3.1. We have selected this mug for its familiarity to the reader as a common household object and because it has certain interesting features. The first feature is that it is entirely one color—black. So, shape is the only visual cue that a human observer would use to decompose the mug. The second feature is its topology. The mug has a genus one topology, which is equivalent to a torus. The handle of the mug creates a loop the makes this a genus one surface. Although most objects have genus zero topology, many objects we encounter with our applications do have non-zero topologies where the mug is just one example. The third feature of the mug is that it has clear minima rule parts. The mug basically consists of a cup, a handle, and a base. To use this mug in our computer vision system, we have created a full computer model of the mug as a triangle mesh using techniques from (Sun and Abidi, 2001). Fig. 3.2 shows examples of this computer mesh model. This mesh is the input to our system. The decomposition of this mesh into a cup, a handle, and a base is the output. In the next section, we clearly specify this input and output in further detail.

3.2   Input and Output

To clarify the goal of our research, we seek to define precisely the input and output to our Minima Rule Algorithm. We begin with an assumption about the underlying surface that our mesh data approximates. We assume that our mesh is a discrete sampling of a piecewise smooth surface. If we denote the piecewise smooth surface as $\check{S}$, then we can denote each part r of $\check{S}$ as $\check{S}_r$ where we define a part in accordance with the minima rule. Each $\check{S}_r$ is also a piecewise smooth surface. We now formally state our problem as follows.

Given a piecewise linear approximation M of a piecewise smooth surface $\check{S} = \bigcup_r \check{S}_r$ where the minima rule defines each part $\check{S}_r$,

Find the corresponding mesh decomposition $M_R = \bigcup_r M_r$ where each M_r approximates the original visual part $\check{S}_r$.

The union operation $\bigcup$ is over the set of R parts where $r = 0, \ldots, (R-1)$. For definitions of $\check{S}$ and M, recall Sec. 2.2.1. Note that $M = (K, V)$ where K is the mesh topology and V is the vertices of the mesh. An example of a mesh for a coffee mug appears in Fig. 3.3(a). The geometry of the mesh is the point cloud V as in Fig. 3.3(b) where we have the set:

$$V = \{v_0, v_1, \ldots, v_{n-1}\} \tag{3.1}$$

and n is the number of vertices in the mesh. The topology K defines the connectivity of the vertices, edges, and faces of the mesh as in Fig. 3.3(c). A precise definition of K is a *simplical 2-complex* (Kinsey, 1993) with the following *k-skeletons*

$$
\begin{aligned}
\text{vertices:} \quad & K^0 = \{0\}, \{1\}, \{2\}, \ldots, \{(n-1)\} \\
\text{edges:} \quad & K^1 = \{0, 1\}, \{1, 2\}, \{0, 2\}, \ldots \\
\text{faces:} \quad & K^2 = \{0, 1, 2\}, \ldots
\end{aligned}
$$

where $K = \bigcup_{k=0}^{2} K^k$. So, $M = (K, V)$ approximates our real world surface $\check{S}$.

We use the term *approximates* to emphasize that measurement error often corrupts the creation of M. Refer to Fig. 3.4 that shows the steps to generate M from $\check{S}$. This figure shows the sampling block where a laser range scanner or some other computer vision sensor samples the real world surfaces of interest. The addition of noise error models measurement error that might occur during the sensing process. The point cloud V is the set of sampled points that are on or near the surface $\check{S}$. The reconstruction block recovers the topology K, or in other words connects the dots, to form the triangle mesh M.

The decomposition of M yields the segmented mesh M_R, which consists of parts M_r where again $r = 0, \ldots, (R-1)$. Refer to Fig. 3.5(a). Each mesh M_r

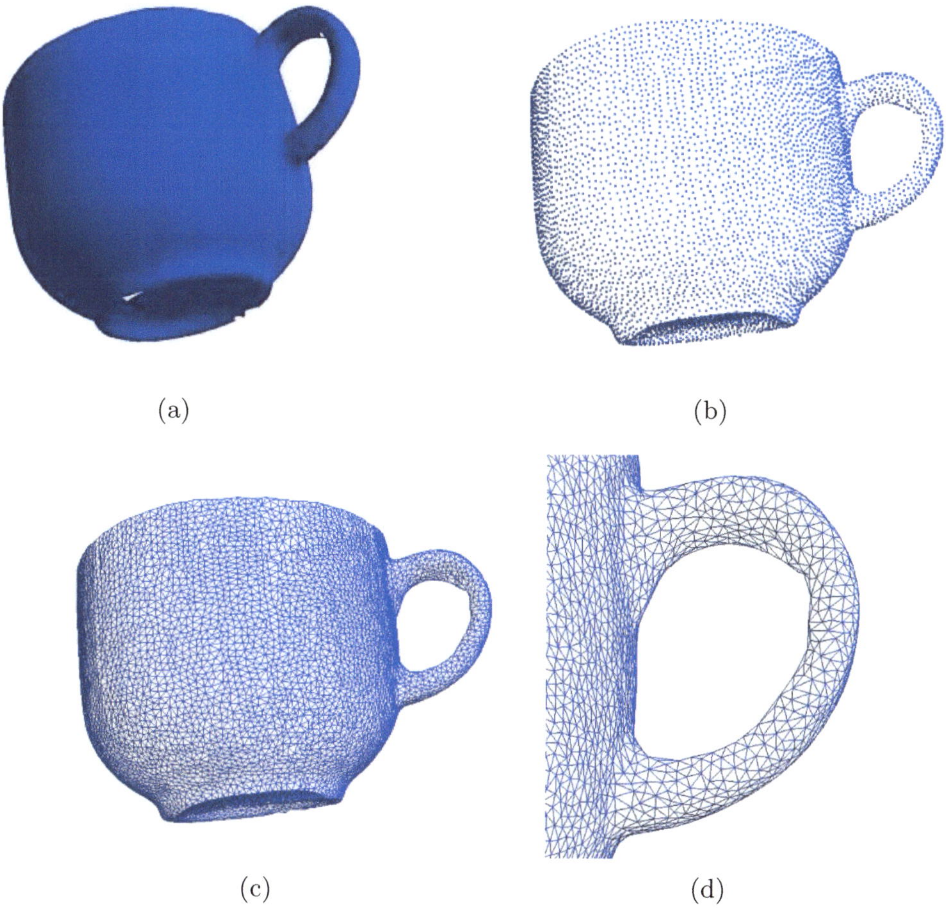

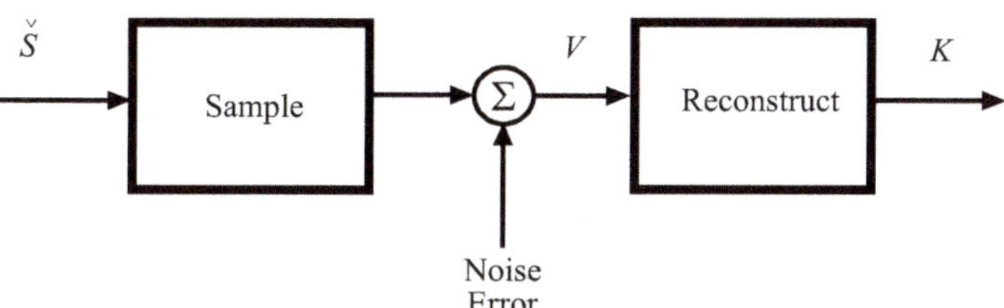

Figure 3.3: We show the input mesh M as (a) A solid rendered model, (b) A point cloud model of just the mesh vertices, and (c) A triangle only model. (d) A zoom view of the handle in (c).

Figure 3.4: This block diagram illustrates the creation of a triangle mesh $M = (K, V)$ from a piecewise smooth surface $\check{S}$.

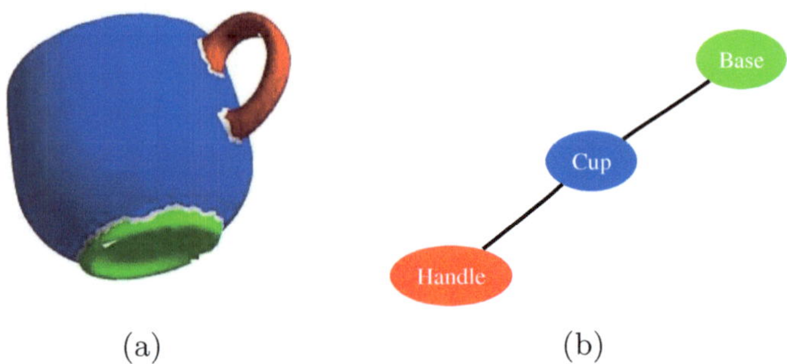

Figure 3.5: We show (a) the output mesh M_R where $R = 3$ and (b) the output graph G.

fits the same definition of a mesh as M such that $M_r = (K_r, V_r)$. Also, the decomposition generates an adjacency graph G, as in Fig. 3.5(b), that represents the interconnection of each part M_r to form M_R. Thus, M is the input, and M_R and G are the output to our Minima Rule Algorithm.

3.3   Algorithm Overview

With the input and output defined, the block diagram in Fig. 3.6 shows the three steps that are necessary to decompose a triangle mesh M into minima rule parts M_R and adjacency graph G. This diagram defines our system and thus our contribution to the state of the art. The first block is the Normal Vector Voting algorithm, which estimates the curvature at each vertex of a mesh. Curvature estimation is fundamental to the minima rule, and Normal Vector Voting is a significant contribution to the start of the art. We develop the theory for this algorithm in Ch. 4. The second block is the Fast Marching Watersheds algorithm for mesh segmentation. This algorithm implements a hill climbing definition of the classic watershed algorithm from image processing to segment our input mesh into the minima rule parts. Ch. 5 presents the details of this algorithm. The final block estimates the quality of the segmentation from the watershed and attempts to improve this quality. The central theme of this block is the Part Saliency Metric, which Ch. 6 addresses. The implementations for these three blocks also include contributions to the state of the art, and each chapter specifically delineates those contributions. Also, each algorithm has either two or three parameters that a user must specify. In general, the Minima Rule Algorithm is not sensitive to these parameters in that a user does not need to tweak each one to gain useful results. Rather, these parameters allow the user to control each stage of the decomposition algorithm to optimize for certain conditions. At the end of each chapter, we present a table of the user parameters for that chapter and specify

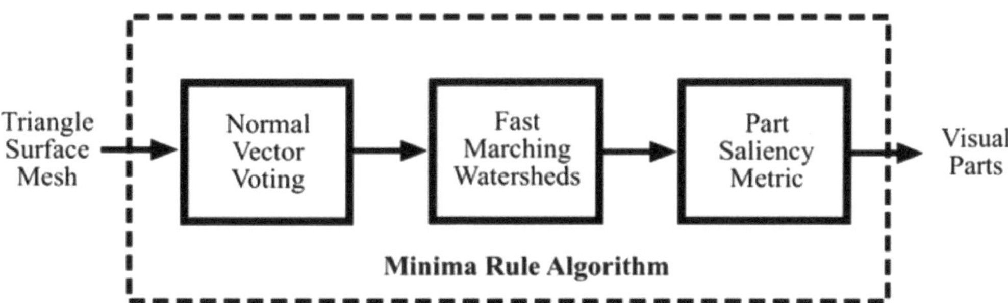

Figure 3.6: Block diagram of the algorithm proposed in this dissertation. The input is a triangle mesh representation that approximates a 3D surface. The output is a set of meshes that are the visual parts of the original surface.

typical values. This concludes our overview of the Minima Rule Algorithm, and we now begin a more in depth look at Normal Vector Voting in the next chapter.

Chapter 4

Curvature Estimation: Normal Vector Voting

Surface curvature plays a key role in tasks such as registration, segmentation, simplification, recognition, and analysis. We find curvature in reverse engineering (Alrashdan et al., 2000; Yang and Lee, 1999), medical visualization (Sander and Zucker, 1986; Sander and Zucker, 1990), and robot navigation (Ferrie et al., 1993) among other applications. For this dissertation, we are interested in curvature as the foundation for our part decomposition algorithm of the minima rule. The importance of curvature is that as a local surface feature it is invariant to rigid transformations and thus serves as a valuable shape description. The major drawback, however, is that it follows from the second derivative of a surface and as such is often difficult to estimate in the discrete world of graphical models. Our goal is to estimate the curvature of a surface from a dense mesh approximation of that surface and in so doing we recognize two key challenges—surface noise and smoothness discontinuities.

First, we consider surface noise. Errors in measurement and registration manifest themselves as noise in the geometry of the mesh. Range imaging in computer vision for example samples the surfaces of a scene and creates point-cloud models. The precision in estimating the position of these points is a function of the sensor mechanics, instrument electronics, surface orientations, and reflective properties. With the variability among these elements, measurement error is inevitable. Additionally, registration error results from reconstruction algorithms that take multiple point clouds as input and attempt to recover the topological relationship among those points relative to the original surface topology. The complexity of aligning the coordinate systems of independent point clouds is a common source of error. Beyond measurement and registration error, isosurface extraction in medical imaging introduces another source of error. Most medical imaging systems generate gridded volume data. Extraction algorithms sift through these grids to create an isosurface mesh. The nature of these algorithms is such that artifacts usually corrupt the output. Although these sources of error listed above are systematic and not necessarily random, we model them as surface noise on the mesh. Filtering and processing often minimize the effects of this noise, but they do not eliminate it entirely.

Second, we consider smoothness discontinuities. Since we assume that our meshes approximate some unknown real-world surface, a question that we must consider is what class of surfaces do we expect. As a practical matter, we restrict ourselves to piecewise-smooth surfaces. This assumption implies that curvature discontinuities are present where two or more smooth surfaces join. Since curvature is singular at such junctions, we must account for these discontinuities. We justify our choice of piecewise-smooth surfaces since most computer vision applications and most medical applications assume a scene consisting of either rigid or non-rigid objects, respectively (Campbell and Flynn, 2001; Flynn and Jain, 1989). Alternatives might include entirely smooth surfaces or piecewise-linear ones, which intuition suggests are not practical models of real-world surfaces. Piecewise smooth is the most appropriate choice but for curvature estimation requires careful consideration at creases.

In this chapter, we describe an algorithm called *Normal Vector Voting* that addresses both of the above issues and robustly estimates curvature for dense triangle meshes. The contributions of Normal Vector Voting are as follows:

- application of geodesic neighborhoods to improve curvature estimation on large dense meshes,

- robust classification of surface orientation to account for curvature singularities at creases and corners, and

- robust estimation of principal directions *and* principal curvatures to overcome surface measurement noise.

The first is the application of geodesic operations to curvature estimation. The dense triangulations of large meshes from computer vision and medical imaging enables geodesic operations to overcome sampling noise and thus to improve the quality of estimation. Another contribution is the crease detection scheme that allows the algorithm to designate a mesh vertex as either on a smooth surface, at a crease junction, or with no preferred orientation. The advantage of this classification is the detection and avoidance of curvature singularities. Finally, a third contribution is the robust estimation of both the principal directions and principal curvatures. Previous methods have demonstrated robust computation of the principal curvatures alone but not the directions (Tang and Medioni, 1999) while others have demonstrated the computation of both but not in a robust manner (Taubin, 1995). Normal Vector Voting bridges the gap between these two algorithms.

In the following sections, we outline the Normal Vector Voting algorithm. We begin in the next section, Sec. 4.1 with a quick overview of Taubin's formulation of curvature estimation for triangle meshes. Then, in Sec. 4.2, we present an overview of the major components of our algorithm. The remaining sections, Secs. 4.3–4.6, discuss each of the components in detail. Finally, we close the chapter in Sec. 4.7 with a few comments.

4.1 Discrete Estimation

By way of introduction, we review the discrete formulation of curvature in accordance with Taubin (Taubin, 1995). Taubin shows that for a point p on a smooth surface the symmetric matrix

$$\mathbf{B}_p = \frac{1}{2\pi} \int_{-\pi}^{\pi} \kappa_p(T_\theta) T_\theta T_\theta^t d\theta \,, \tag{4.1}$$

with superscript t denoting transpose and T_θ is a column vector as defined in Sec. 2.2.1, has eigenvectors that are equivalent to the principal directions $\{T_1, T_2\}$ and eigenvalues that are related by a fixed homogeneous linear transformation to the principal curvatures as

$$\begin{aligned} \kappa_p^1 &= 3b_p^1 - b_p^2 \\ \kappa_p^2 &= 3b_p^2 - b_p^1 \end{aligned} \tag{4.2}$$

where b_p^1 and b_p^2 are the eigenvalues of $\mathbf{B}_p$ associated with T_1 and T_2, respectively. The third eigenvalue is zero and corresponds to the eigenvector equal to the surface normal at p. For a vertex v on a discrete mesh, Taubin approximates (4.1) as

$$\tilde{\mathbf{B}}_v = \frac{1}{2\pi} \sum w_i \kappa_i T_i T_i^t \tag{4.3}$$

for a finite set of directions T_i in the tangent plane of v. The weight w_i is the discrete integration step and has the constraint $\sum w_i = 2\pi$. Taubin's algorithm computes $\tilde{\mathbf{B}}_v$ for a vertex on a mesh and then decomposes the matrix with a Householder transformation and a Givens rotation. The resulting eigenvectors and eigenvalues lead to the principal directions and principal curvatures via (4.2).

The question at hand is how do we estimate κ_i and T_i in (4.3). Taubin employs a truncated Laurent series to approximate these values, but this approach is not robust. Tang and Medioni (Tang and Medioni, 1999) suggest a more robust solution. Building on these algorithms, we have developed Normal Vector Voting

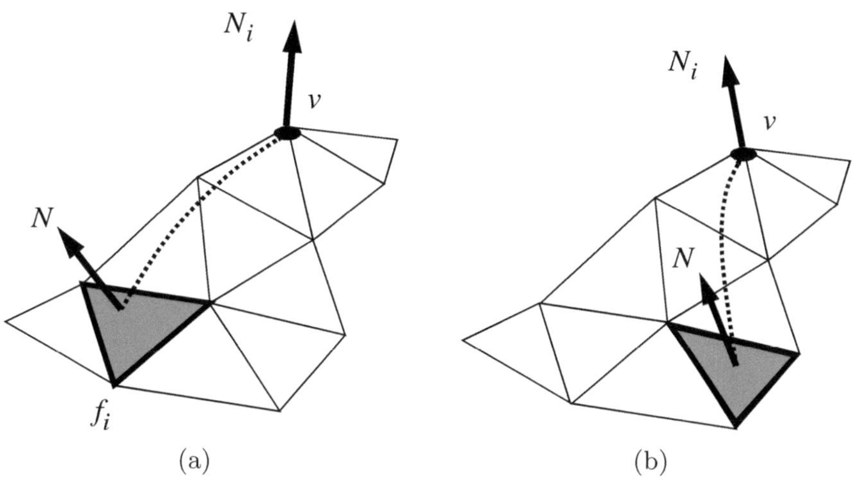

Figure 4.1: A triangle with normal N generates vote N_i for the orientation of the surface at v. The circular arc follows the perceptual continuity constraints.

as a robust method to estimate the individual κ_i and T_i and thus the principal directions and principal curvatures for a vertex on a mesh. We now take an in-depth look at this algorithm.

4.2 Algorithm Overview

Normal Vector Voting is a two-pass algorithm. For the first pass through a mesh, we estimate the normal vector orientation for each vertex. For the second pass, we estimate curvature. For normal vector estimation, the basic idea is to select a surface region around a vertex. A user-specified distance bounds this region in terms of geodesic distance from the vertex where the vertex is the center of the geodesic patch. Each triangle in this patch—or geodesic neighborhood—*votes* at that center vertex in order to estimate the orientation of that vertex. Note the simple example in Fig. 4.1. Here, triangle f_i in the mesh neighborhood M_v of vertex v has a normal N that generates a normal vote N_i at v. We collect these votes in a covariance matrix and decompose this matrix using eigen analysis. The eigenvectors and eigenvalues estimate the orientation for v where the orientation is either a surface normal N_v, a crease tangent T_v, or a null vector for no orientation. We illustrate this sequence of events in Fig. 4.2. With a few slight modifications, this same sequence estimates the curvature at v for the second pass. The algorithm in Alg. 4.3 demonstrates both passes and the equations in this algorithm are in the following sections.

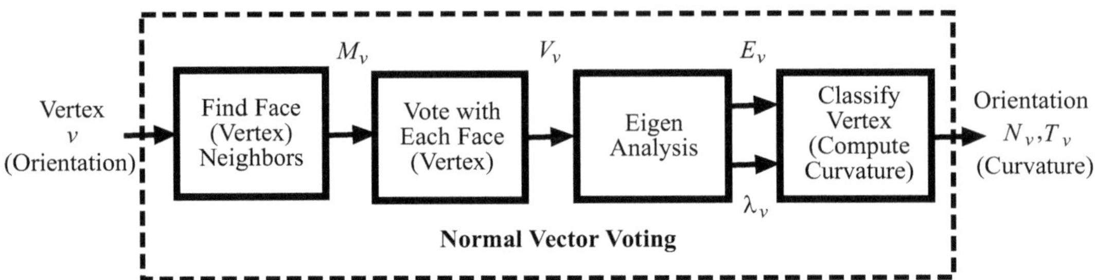

Figure 4.2: This block diagram shows the flow of the Normal Vector Voting process for a single vertex. With slight modifications, the same diagram applies for the estimation of the surface normal for a vertex and of the surface curvature for a vertex.

4.3 Geodesic Neighborhood

The first step in Normal Vector Voting for both the first or second pass is to find the triangles or vertices that are close in a geodesic sense to the vertex of interest. The *geodesic neighborhood* problem, which follows the *discrete geodesic problem* (Mitchell et al., 1987), is to find the m triangles that are within a user-specified distance of the vertex. The key is that the distance is not the Euclidean distance but rather the shortest geodesic distance along the surface of the mesh. As noted in the literature this problem closely resembles the shortest path problem for a graph, which the Dijkstra algorithm (Dijkstra, 1959) solves. The difference is that the shortest path along the edges and nodes of a graph is not necessarily equivalent to the shortest geodesic path over the surface, which includes the triangle interiors and not just the edges. Kimmel and Sethian (Kimmel and Sethian, 1998) present an elegant algorithm, called Fast Marching, that solves this problem in $O(m \log m)$ time and Sun and Abidi (Sun and Abidi, 2001) propose a simplification that is readily adaptable to our domain. The time complexity is important when compared to such Euclidean algorithms as k-d trees that require $O(m+3n^{\frac{2}{3}})$ time where n is the number of triangles in the whole mesh (Weiss, 1999). Also, k-d trees require additional $O(n)$ storage space beyond the current space required for the mesh, itself. The Kimmel and Sethian algorithm requires no additional storage.

As a brief aside, we distinguish the Kimmel and Sethian definition of shortest geodesics from the straightest geodesics of Polthier and Schmies (Polthier and Schmies, 1998; Polthier and Schmies, 1999). The latter insures the uniqueness of a geodesic path on a polyhedral surface. Since we are only concerned with a neighborhood and not the actual path, this trait is not crucial to our problem. We are interested in shortest and not straightest geodesics.

Fig. 4.4 shows three different sizes of geodesic neighborhoods. The smallest neighborhood in this figure consists of just the immediate triangles adjacent to the vertex of interest. We also refer to this simple patch as the *umbrella neighborhood* since it figuratively resembles a collapsible rain umbrella with a canopy of triangles mounted on its central vertex. Another common term for the umbrella is the *one-ring neighborhood*. Subdivision surfaces refer to the k-ring neighborhood of a vertex where a ring is the set of triangles within k edges of the vertex. A k-ring is a topological neighborhood definition. We will compare k-ring topological neighborhoods and geodesic ones shortly. Most algorithms that require a neighborhood usually work with the umbrella patch. Taubin's curvature algorithm (Taubin, 1995), for example, employs such a neighborhood. Our research has shown—and hopefully Normal Vector Voting demonstrates—that significant advantages arise if we enlarge beyond the umbrella neighborhood. The choice of the neighborhood size depends on the application but in our context we identify feature size, noise level, and sampling density as the variables that dictate this choice. These variables have competing interests and thus tradeoffs among them exist. For example, a mesh that contains small features requires a small neighborhood to preserve those features. If that same mesh, however, has significant noise corruption, then a large neighborhood is necessary to smooth out the noise. Similar arguments follow for sampling density. Because of these competing issues, we have defined the neighborhood size as a user-specified parameter. The user simply specifies the geodesic distance that bounds the neighborhood. Actually, the user specifies an integer multiple of the average length of the triangle edges in the mesh to eliminate scaling issues. As a result we define the *k-geodesic neighborhood* as the neighborhood with a geodesic boundary that is k times the average edge length. We derive the term k-geodesic in the spirit of the k-ring designation.

A question that arises is what is the benefit of a k-geodesic neighborhood over a k-ring neighborhood. In particular does the computational burden to find a k-geodesic overshadow that for a k-ring. For insight, we outline the Kimmel and Sethian's Fast Marching algorithm. We begin at the vertex of interest and simply walk outward to the one-ring vertices. We use the equations in (Sun and Abidi, 2001) to estimate the shortest geodesic distance back to the center vertex from each one-ring vertex. We then place these vertices on a heap with their distance as a key into the heap. We now walk to the closest one by removing the top vertex from the heap—the vertex with the shortest distance. We compute the geodesic distances for each of its one-ring vertices back to the original center and place them on the heap. The algorithm again removes the closest vertex from the heap and repeats. We see that this walking algorithm is very similar to a k-ring neighborhood algorithm. The only differences are the computation of the geodesic distance for each vertex and the priority of the walk driven by the heap. A ring algorithm would simply prioritize the walk as a function of the current ring state where the walk proceeds after extending the neighborhood to a complete ring. The geodesic neighborhood algorithm walks towards the closest vertex regardless of how close or how far that vertex is in terms of a ring.

Although this priority walking does require some additional computation, the benefits, especially for small neighborhoods, far outweigh this minor cost as the following examples illustrate. For the regular triangulations that we see with range images and isosurface algorithms, we often find a bias of a one-ring neighborhood as shown in Fig. 4.5(a). The thick dark circles in this figure represent k-geodesic boundaries while the alternating bands of white and shaded triangles represent the various k-ring neighborhoods. Inside the first dark circle, the small one-ring neighborhood has a bias towards the right side of the figure. A one-geodesic neighborhood, however, defined by the first circle includes the one-ring triangles and the triangles labeled with an "A" where we consider a triangle to be in the one-geodesic neighborhood if each of its three vertices fall within the region bounded by the circle. These additional triangles balance the support of the neighborhood and eliminate the unintended bias of a simple one-ring. Similarly, the white two-ring also has a bias towards the right where the inclusion of the "B" triangles offers a more balanced two-geodesic neighborhood. Similarly, triangulations that are not quite as regular such as the one in Fig. 4.5(b) have a bias as well since triangle sizes and configurations have more variation. We again use the "A" and "B" labels to illustrate the comparison. For curvature estimation our experience suggests that the balanced support of geodesic neighborhoods are a significant benefit especially in the context of the discrete (4.3) where adequate directional sampling around a vertex is important.

Since Normal Vector Voting is a neighborhood-oriented operation with a geodesic definition, we label it as a *geodesic operation*. We use this label in the spirit of the so-called *mask operations* (Gonzalez and Woods, 1993) from image processing. With mask operations, we analyze an image pixel in terms of its own gray level and of the gray levels of its neighbors. We often specify the mask neighborhood as $k \times k$ where k is the width in pixels of the mask centered at the pixel of interest. We see that our geodesic operation is similar except we specify the k-geodesic neighborhood of interest. We now have a geodesic neighborhood M_v. Our next step is to vote and determine the orientation of v or curvature at v depending on if we are in the first or second pass, respectively.

4.4   Vote Collection

For the first pass, the next block in the diagram of Fig. 4.2 involves the voting of the triangle faces $f_i \in M_v$ at the vertex v. We must address two questions:

- How does face f_i *cast* a vote?

- How does vertex v *collect* these votes?

To answer these questions, we are inspired by Tensor Voting (Medioni et al., 2000). Tensor Voting is a computational framework that infers structures such as boundaries and surfaces from unstructured, sparse, and often noisy 3D point clouds. This framework employs perceptual constraints from theories of human vision and subsequent definitions of tensor voting fields to extract structure. The implementation of tensor voting requires a discrete voxel representation of space where input points *cast* votes and voxels *collect* votes in the context of tensor algebra. Medioni et al. suggest a system of tensor voting fields in conjunction with coordinate frame rotations and convolution-like operations with those fields to encode local geometric structure at each voxel. After the voting process, they extract salient global structures such as surfaces, curves, and junctions by sifting through these voxels with a level-crossing detection algorithm. Although the application of this approach to triangle meshes is possible, we reformulate the Tensor Voting notions and propose a more natural vector framework for triangle meshes. We use the same perceptual continuity constraints (Andreson, 1995) as Medioni et al. to govern vote casting, but we define a more appropriate vector geometry instead of the tensor voting fields. (Recall Fig. 2.2 for an illustration of perceptual constraints.) For vote collection, we use a covariance matrix similar to the quadric error matrix in (Heckbert and Garland, 1999), which has a direct relationship to the tensor algebra in Medioni et al. Interestingly, Heckbert and Garland (Heckbert and Garland, 1999) show a relationship between this covariance matrix—in the limit—to surface curvature. We however use this matrix to estimate orientation and later follow Taubin's approach to estimate curvature. We have found that, in the presence of surface noise, this two-step approach is more robust than directly extracting curvature from the covariance matrix.

4.4.1 Casting Votes

We first consider how a triangle f_i generates a normal vote N_i at vertex v. A couple of approaches are possible. For example, one method is to set N_i at v equal to the normal N of the plane that contains f_i as in Fig. 4.6(a), so that $N_i = N$. This simple method works well with low curvature surfaces but leads to significant error as curvature increases. With some insight, we see that an improvement to this method is to fit a smooth curve from f_i to v and allow the normal vote to follow the curve. The perceptual continuity constraint (Andreson, 1995) suggests that the most appropriate curve is a circular arc with shortest arc length. Following this argument, we construct the geometry in Fig. 4.6(b). For the shortest arc-length circle, the curve must lie entirely in the plane Π_i that contains the triangle normal N—rooted at centroid c_i of triangle f_i—and the vertex v. We can compute θ_i in the figure as

$$\cos \theta_i = -\frac{N^t vc_i}{\|vc_i\|}$$

where $vc_i = c_i - v$ and $0 \leq \theta_i \leq \pi$. This equation leads to the normal vote

$$N_i = N + 2\cos\theta_i \frac{vc_i}{\|vc_i\|} \, . \tag{4.4}$$

4.4.2 Collecting Votes

We next address how v collects the N_i votes from each $f_i \in M_v$. One possibility is as a weighted vector sum $\sum w_i N_i$. This approach is a common method for normal vector estimation at vertices in a triangle mesh, typically with an umbrella neighborhood. The limitation of this scheme is that normals with opposite orientation annihilate each other ($N_j = -N_i$, $N_j + N_i = \vec{0}$), and we therefore lose variance information. This situation occurs near crease and corner discontinuities in particular. As an alternative, we represent N_i as a covariance matrix $\mathbf{V}_i = N_i N_i^t$ and collect votes as a weighted matrix sum $\mathbf{V}_v$ with

$$\mathbf{V}_v = \sum w_i \mathbf{V}_i = \sum w_i N_i N_i^t \tag{4.5}$$

where the summation is over the M_v neighborhood. Unfortunately, the downside is that we now lose absolute sign orientation. The covariance matrix $\mathbf{V}_i$ does not designate which side of the mesh is the outside of the surface. Consider $N_j = -N_i$. How do we distinguish $N_i N_i^t = N_j N_j^t$? The benefit is that these votes no longer annihilate each other. We can track the variance of the votes. This capability outweighs the loss of the absolute sign information since the variance allows us to draw conclusions about the relative orientation of the vertex. We will see in the next section that eigen analysis of this variance leads to an interesting classification scheme for the vertex. With regard to the absolute sign, we should be able to recover this information with a simple *ad hoc* algorithm such as a quick averaging of the umbrella neighbors. Only under a pathological case does such an approach fail. So we can readily overcome the sign problem.

4.4.3 Weighting Votes

Our final issue to address is the weighting term w_i. Two factors effect this term: surface area of f_i and geodesic distance, g_i, of c_i from v. Naturally, a triangle closer to v should have a stronger vote than a triangle farther away while a larger one should also have a stronger vote than a smaller one. We choose an exponential decay to reflect these notions

$$w_i = \frac{A_i}{A_{max}} \exp\left(-\frac{g_i}{\sigma_k}\right) \tag{4.6}$$

where σ_k controls the rate of decay, A_i is the area of f_i, and A_{max} is the area of the largest triangle in the entire mesh. In practice, the value, σ_k, is a function of the maximum geodesic distance, g_m, that the user specifies as a k multiple of the average edge length in the mesh. We define this value as

$$g_m = 3\sigma_k = kl_{ave} \tag{4.7}$$

where l_{ave} is the average length of the triangle edges in the mesh. Votes beyond the neighborhood g_m have negligible influence and can be ignored.

The above equations lead to a covariance matrix $\mathbf{V}_v$ for each vertex v in the mesh. This matrix represents the variance of the votes in the geodesic neighborhood M_v around v. In the next section, we use eigen analysis to investigate the orientation of v using this variance information.

4.5   Orientation Classification

While still in the first pass of the algorithm, the third step in Fig. 4.2 is to decompose $\mathbf{V}_v$ using eigen analysis and then to classify v. Since $\mathbf{V}_v$ is a symmetric semi-definite matrix, eigen decomposition generates real eigenvalues $\lambda_1 \geq \lambda_2 \geq \lambda_3 \geq 0$ with corresponding eigenvectors E_1, E_2, and E_3. We can visualize this eigensystem as in Fig. 4.7(a). The eigenvalues, and thus the shape of the eigen ellipsoid, yield insight into the vote agreement within $\mathbf{V}_v$. Figs. 4.7(b)–4.7(d) shows three variations of the eigenvalues and how we might interpret these variations.

In Tensor Voting, Medioni et al. (Medioni et al., 2000) define saliency maps over the entire voxel space with eigenvalues from their tensors. They then use an extremal search algorithm to extract salient global structures from these map definitions. The saliency maps use the following relationships for their tensor eigenvalues:

$$\begin{aligned}
\mathcal{S}_s &= \lambda_1 - \lambda_2 \,, &&\text{surface patch saliency;} \\
\mathcal{S}_c &= \lambda_2 - \lambda_3 \,, &&\text{crease junction saliency; and} \\
\mathcal{S}_n &= \lambda_3 \,, &&\text{no preferred orientation saliency.}
\end{aligned} \qquad (4.8)$$

Since we seek to classify the preferred orientation of a vertex using vector algebra as opposed to the global structure through a voxel using tensor algebra, we take a different approach and do not employ a search algorithm to sort through voxels. For our vector voting, we propose the following vertex classification scheme for the eigenvalues of $\mathbf{V}_v$ at each vertex:

$$\max\{\mathcal{S}_s, \varepsilon\mathcal{S}_c, \varepsilon\eta\mathcal{S}_n\} = \begin{cases} \mathcal{S}_s: & \text{surface patch with normal } N_v = E_1 \\ \varepsilon\mathcal{S}_c: & \text{crease junction with tangent } T_v = E_3 \\ \varepsilon\eta\mathcal{S}_n: & \text{no preferred orientation} \end{cases} \qquad (4.9)$$

where $0 \leq \varepsilon < \infty$ and $0 \leq \eta < \infty$ are constants that control the relative significance of the saliency measures. These constants are not user parameters since they are fixed for a given system. When we design a system however we need to carefully select these constants to balance noise tolerance and crease detection.

We demonstrate the design impact of ε and η with examples. First, consider one extreme where we design $\varepsilon = \eta = 0$. This system always classifies a vertex as a surface patch regardless of any corners or creases in the original piecewise-smooth surface. Consider

$$\lim_{\varepsilon,\eta\to0} \max\{\mathcal{S}_s, \varepsilon\mathcal{S}_c, \varepsilon\eta\mathcal{S}_n\} = \mathcal{S}_s \,.$$

This design associates a surface normal with each vertex even if the vertex is a sample of a crease or a corner. Thus, the design does not detect curvature discontinuities. This approach is very similar to a normal estimation algorithm that averages the triangle normals of a a one-ring neighborhood for a vertex. Another design extreme lets $\varepsilon, \eta \to \infty$. Such a system never classifies a vertex as a surface patch regardless of smoothness and instead classifies each vertex as a corner. This design never assigns surface normals to vertices. The third extreme is a design where $\varepsilon \to \infty$ and $\eta = 0$. As we might expect, this system always classifies a vertex as a crease and associates a tangent vector with the vertex. Although the first design may have some use, the latter two designs have little practical use, but they do illustrate the choice of the constants (ε, η). When designing a system, we fix ε to discriminate the types of creases that we expect in the piecewise-smooth surfaces and η to discriminate the amount of surface noise that we wish to tolerate in our sensors. extrema of the classification constants.

Table 4.1: Extrema values for classification constants.

A	B
C	D

For a system design, we need to decide how much noise we can tolerate and what crease angles we need to detect. If we choose to detect small creases, we reduce the overall robustness of the system. On the other hand, less tolerance to small creases allows more tolerance to noise. The constant ε controls these design considerations. In our experiments, the system $\varepsilon = 2$ offers a balanced compromise of detecting creases and allowing variation. As a rule of thumb, we have the following equation

$$\tan\frac{\phi}{2} = \sqrt{\frac{1}{\varepsilon + 1}} \tag{4.10}$$

where ϕ is the minimum dihedral crease angle that the system can detect. We illustrate examples of crease angles in Fig. 4.8. We emphasize that this angle is not for edges in the mesh but for the creases of the original piecewise-smooth surface. So, for $\varepsilon = 2$, we detect creases in the original surface with $\phi \geq \frac{\pi}{3}$. If however that surface has creases with $\phi < \frac{\pi}{3}$, the system classifies vertices that are samples near these creases as surface patches, but the benefit is robustness to noise. Following a similar example and argument, we can see that $\eta = 2$ also offers a balance between noise and crease detection. We can formulate the following equation

$$\tan\frac{\psi}{2} = \sqrt{\frac{1}{\eta + 1}} \tag{4.11}$$

where ψ is the angle of variation between a crease decision and a no-preferred-orientation decision relative to the eigen analysis of $\mathbf{V}_v$ from the previous section. With most systems, we suggest $\varepsilon = 2$ and $\eta = 2$.

With our first pass through the mesh, this classification estimates the normals for each vertex on a surface patch and detects each vertex along a crease discontinuity. Using this information with extensions to Taubin's algorithm, we discuss in the next section how a second pass through the mesh generates estimates for the curvature at each vertex.

4.6 Curvature Estimation

Our second pass through the mesh follows the same sequence as in the first pass. Recall Fig. 4.2. This time, however, we use the normal estimates from the previous section to estimate the curvature at each vertex. We again use a geodesic neighborhood M_v around each vertex but for this pass we are interested in the vertices $v_i \in M_v$, and not the triangles, in this neighborhood. Each vertex v_i votes at the center vertex v where we collect the votes in a matrix $\tilde{\mathbf{B}}_v$ from Eqn. (4.3). We decompose this matrix with eigen analysis and use the subsequent eigenvectors and eigenvalues to estimate the directions T_1 and T_2 and principal curvatures κ_v^1 and κ_v^2 at v with the linear transformations in Eqn. (4.2). We now specify the weights w_i, tangent directions T_i, and normal curvatures κ_i in the matrix sum of $\tilde{\mathbf{B}}_v$.

We begin with the weights since they are the simpliest terms to define. As with the first pass, we use the same decay function in Eqn. (4.6) except that we remove the area components A_i and A_{max}. Also, we constrain $\sum w_i = 2\pi$ for all the weights around v. This constraint is necessary to maintain translation invariance among the votes. Again, the decay function places more emphasis on votes that are closer to v than ones that are farther away.

We use the geometry in Fig. 4.6(c) to define the tangent directions T_i of each vote. The figure demonstrates that we project the vector from v_i to v into the tangent plane of v and normalize the result. The following equation is more precise

$$T_i = \frac{t_i}{\|t_i\|} \ , \ t_i = vv_i - (N_v^t vv_i)N_v \tag{4.12}$$

where $vv_i = v_i - v$. This direction is for any vertex v_i in the geodesic neighborhood of v and not just the umbrella neighbors as in Taubin's algorithm. The normal N_v is the estimate from the previous section.

Last we consider the normal curvatures κ_i. We propose a discrete definition using the changes in turning angle ϑ_i and in arc length s where

$$\kappa_i = \frac{\Delta\vartheta_i}{\Delta s} \ . \tag{4.13}$$

An important consideration is that we properly define the turning angle for the normal curve and not just the curve connecting v_i and v. The change in the turning angle describes the change in the normal vector as we move along the curve. To this end, we project the normal estimate N_{v_i} at v_i into the plane Π_i that contains N_v—rooted at v—and v_i as

$$n_i = N_{v_i} - (P_i^t N_{v_i})P_i \tag{4.14}$$

where n_i is the projection and $P_i = N_v \times T_i$ defines the plane that contains the normal curve. The turning angle thus becomes

$$\cos(\Delta\vartheta_i) = \frac{N_v^t n_i}{\|n_i\|} \ . \tag{4.15}$$

The sign of κ_i is the same as the sign of $T_i^t n_i$. Finally, the change in arc length is simply the geodesic distance between the two vertices

$$\Delta s = g_i \ . \tag{4.16}$$

We estimate g_i from the geodesic neighborhood algorithm discussed in Sec. 4.3. For a vertex v, we collect the curvature votes from the equations above into the matrix $\tilde{\mathbf{B}}_v$ of Eqn. (4.3), and eigen decomposition leads to the principal directions T_1 and T_2 and principal curvatures κ_v^1 and κ_v^2 from the relation in Eqn. (4.2).

4.7 Remarks

We have reached our goal. After the second pass, we have an estimate for the curvature at each vertex. We discuss a few caveats, however. First we only compute surface curvature if we classify a vertex as a surface patch. If a vertex has no preferred orientation, surface curvature is meaningless. A vertex on a crease on the other hand is a little different since we can estimate the curvature in the direction of the crease. With slight modifications of the above equations, we can generate a tangential curvature estimate. The other principal curvature, which is orthogonal to this one, is infinite, but we can estimate the cusp angle across the crease as either $\phi = 2\arctan\sqrt{\frac{\lambda_1}{\lambda_2}}$ or $(\pi - 2\phi)$ where λ_1 and λ_2 are the eigenvalues from the first pass through the mesh in Sec. 4.5. The choice for the angle depends on the absolute sign information that our *ad hoc* umbrella method resolves. We further note that our classification scheme does not enforce crease continuity, i.e. topologically link crease vertices. If such topological connectivity is important, we suggest morphological operations (Rössl et al., 2000) or watershed methods (Mangan and Whitaker, 1999). The final caveat relates to the neighborhood definition. The neighborhood algorithm is a fast marching method that begins at the vertex of interest v as the center and marches out to form the neighborhood M_v. For curvature estimation as the algorithm marches outward, we check the classification of the current vertex and only proceed if it is a surface patch. This qualification does not allow the marching algorithm to cross crease discontinuities and thus

Table 4.2: Parameters for Normal Vector Voting Algorithm.

Parameter	Range	Equation	Typical Value	Comments
$k,\ \sigma_k$	$0 < k < \infty$	(4.7)	$k = 3$	The k-geodesic neighborhood accounts for surface noise. This parameter is a balance between noise robustness and feature preservation.
ε	$0 \leq \varepsilon < \infty$	(4.10)	$\varepsilon = 2$	Determines the possible creases features that we can detect. This variable is more of a system design parameter than a user one.
η	$0 \leq \eta < \infty$	(4.11)	$\eta = 2$	Determines the level of noise suppression. As with ε, this variable is also a system design parameter.

restricts M_v to the same smooth patch as v. This approach improves the curvature estimate since vertices on the other side of a discontinuity do not corrupt the estimation.

In this chapter, we have noted three parameters that control the Normal Vector Voting algorithm. As a reference, we summarize these parameters in Table 4.2. Recall that the only real user parameter is k and the other two variables, ε and η, are more system design parameters. We make this distinction as the user typically adjusts k to increase the accuracy of the curvature estimates relative to the surface noise of a particular mesh or to the relative size of features in the mesh. The user, on the other hand, should probably not adjust ε and η. We should fix these variables for a particular implementation.

This Normal Vector Voting algorithm, as presented in this chapter, serves as the curvature estimation algorithm for the part decomposition algorithm in this dissertation. In the next chapter, we present a mesh segmentation algorithm, which is the next step in our development of the decomposition theory. The output of Normal Vector Voting serves as the input for the algorithm in the next chapter.

input : A triangle surface mesh M, k-geodesic boundary, and (ε, η) for crease detection

output: Four arrays for principal curvatures and directions, κ_{max}, κ_{min}, T_{max}, T_{min}

foreach *vertex v in M* **do**

    $M_v \leftarrow$ `GeodesicFaceNeighbrhood(` v, k`)`;

    initialize matrix V_v;

    foreach *triangle face f_i in M_v* **do**

        $N_i \leftarrow$ `NormalVectorVote(` v, f_i`)`;

        $w_i \leftarrow$ `ComputeWeight(` v, f_i`)`;

        $V_v \leftarrow V_v + w_i$ `InnerProduct(` N_i, N_i`)`;

    end E_v, λ_v

    $\leftarrow$ `EigenAnalysis(` V_v`)`;

    $(\mathcal{S}_s, \mathcal{S}_c, \mathcal{S}_n) \leftarrow$ `ComputeSalience(`λ_v`)`;

    switch max($\mathcal{S}_s$, $\varepsilon\mathcal{S}_c$, $\varepsilon\eta\mathcal{S}_n$) **do**

        case $\mathcal{S}_s$ **do** surface patch $N_v \leftarrow E_{v,1}$;

        ;

        case $\varepsilon\mathcal{S}_c$ **do** crease junction $T_v \leftarrow E_{v,3}$;

        ;

        case $\varepsilon\eta\mathcal{S}_n$ **do** no preferred orientation;

        ;

    end

end

foreach *vertex v in M that is a surface patch with N_v* **do**

    $M_v \leftarrow$ `GeodesicVertexNeighbrhood(` v, k`)`;

    initialize matrix K_v;

    foreach *vertex v_i in M_v that is a surface patch with N_{v_i}* **do**

        $T_i \leftarrow$ `TangentVectorVote(` N_v, v_i`)`;

        $k_i \leftarrow$ `ComputeNormalCurvature(` N_v, N_{v_i}, T_i`)`;

        $K_v \leftarrow K_v + k_i$ `InnerProduct(` T_i, T_i`)`;

    end E_v, λ_v

    $\leftarrow$ `EigenAnalysis(` K_v`)`;

    $T_{max}[v] \leftarrow E_{v,1}$;

    $T_{min}[v] \leftarrow E_{v,2}$;

    $\kappa_{max}[v] \leftarrow 3\lambda_{v,1} - \lambda_{v,2}$;

    $\kappa_{min}[v] \leftarrow 3\lambda_{v,2} - \lambda_{v,1}$;

end

Figure 4.3: Normal Vector Voting Algorithm for a surface mesh data structure. Note the algorithm is a two-pass algorithm through the vertices of the mesh. The input values (ε, η) are actually design constants and not user variables like k. The user adjusts k to account for noise, sampling, and features in the mesh. However, we fix (ε, η) to detect specific crease features. The user does not need to adjust these values for different meshes.

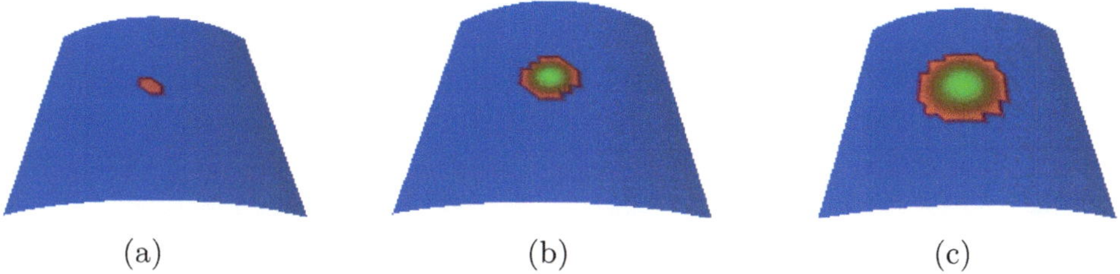

Figure 4.4: Examples of different sizes of geodesic neighborhoods for the same vertex on a cylindrical surface mesh. (a), (b), and (c) have one-, three-, and five-geodesic neighborhoods, respectively.

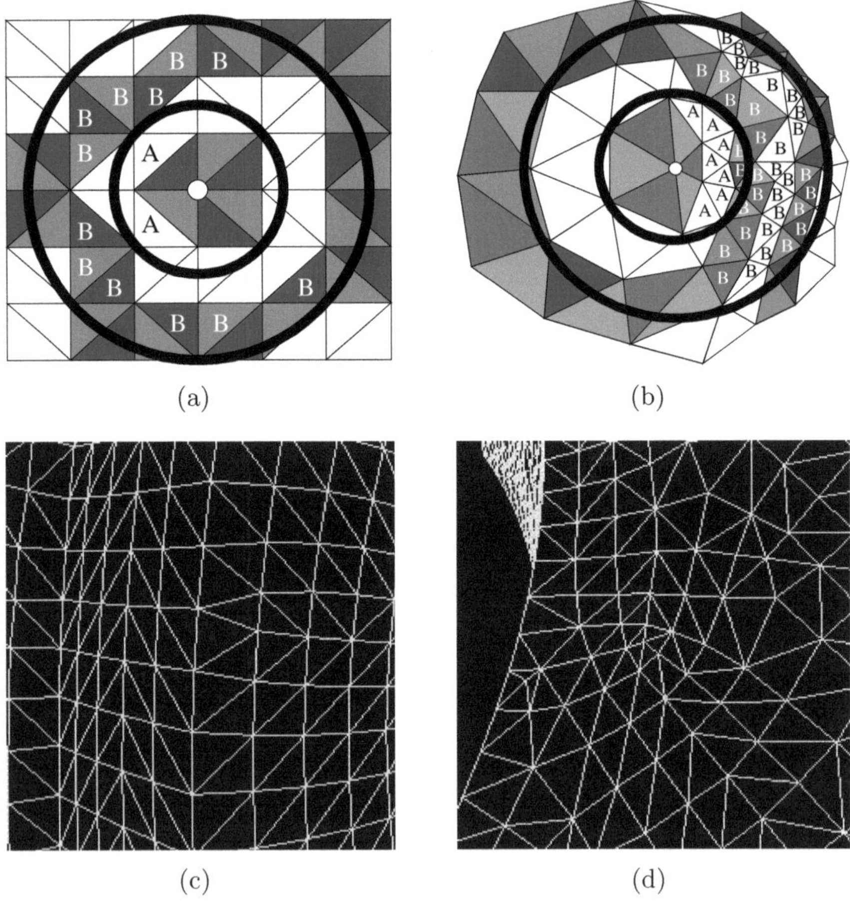

Figure 4.5: Examples of geodesic neighborhoods versus ring neighborhoods. The alternating rings of shaded and white triangles depict the various rings. The thick circles bound geodesic neighborhoods. (a) The "A" triangles distinguish a one-ring neighborhood from a one-geodesic neighborhood. The "B" triangles show a similar difference for a two-ring and a two-geodesic. (b) When triangulations are not as regular, we often see more exaggerated differences. (c,d) Triangulations that exhibit the behavior depicted in (a) and (b).

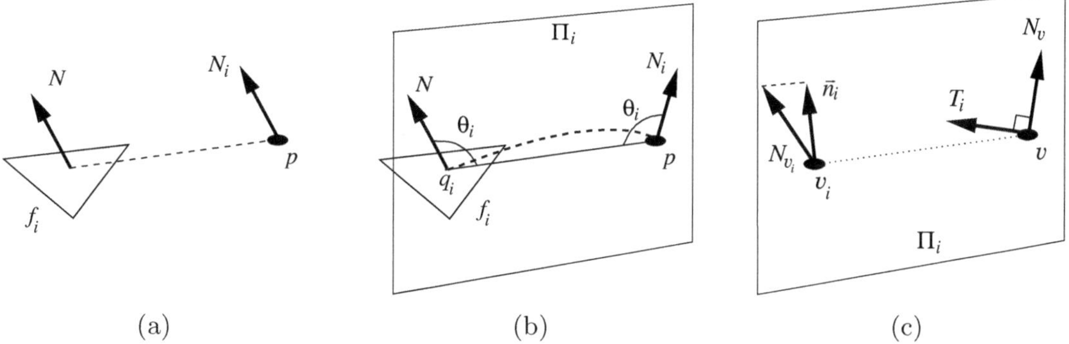

Figure 4.6: Geometries for Normal Vector Voting. (a) A simple translation of N ($N_i = N$). (b) A slightly more sophisticated vote where we rotate N by $(2\theta_i - \pi)$ in plane Π_i. (c) Curvature estimation geometry where N_v, n_i, and T_i lie in the plane Π_i and n_i is the projection of N_{v_i} onto the plane.

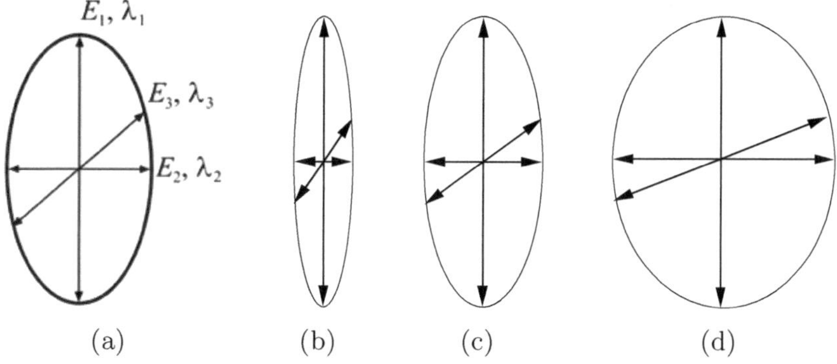

Figure 4.7: These ellipsoids depict possible variations in the eigenvalues. We interpret these variations of $\mathbf{V}_v$ as the orientation saliency of the neighborhood around v. (a) Covariance matrix $\mathbf{V}_v$ where the eigenvalues λ_1, λ_2, and λ_3 define the shape. (b) Surface patch where λ_1 is much larger than λ_2, and λ_3. (c) Crease discontinuity where λ_1, and λ_2 are similar in value but larger than λ_3. (d) Patch with no preferred orientation where each eigenvalue is similar in value.

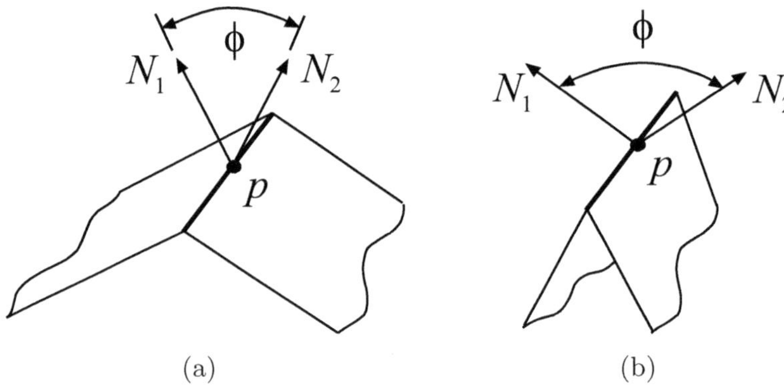

Figure 4.8: These examples illustrate different dihedral angles for creases on piecewise-smooth surfaces. The normal vectors N_1 and N_2 are for the respective smooth patches adjacent to the crease.

Chapter 5

Mesh Segmentation: Fast Marching Watersheds

A triangle mesh is simply a collection of triangles and vertices that approximate a 3D surface. Although this representation is useful for visualization on a computer, this low-level description is often inadequate for other tasks such as object recognition, scene understanding, and feature modeling. We, as humans, can readily observe a 3D mesh rendered on a computer screen and quickly infer higher-level descriptions such as the handle on a mug, but to the computer, the mesh is nothing more than a jumbled pile of triangles and vertices. A computer inherently has no higher notion of the interrelationship of the pile. Higher-level descriptions are necessary through appropriate data structures. One way to impose such descriptions is through mesh segmentation.

Mesh segmentation refers to the partitioning of a mesh into a set of groups or regions that cover the mesh. We emphasize the words *group* and *region* to distinguish the two types of segmentation. A group segmentation clusters vertices and triangles without regard to their topological relationship while a region segmentation classifies topologically connected vertices and triangles. By analogy, if we think of people as mesh vertices and triangles, group algorithms would segment people according to characteristics such as tall, short, skinny, blonde, and brunette while region algorithms would segment people relative to their physical location such as city dwellers, country folks, mountaineers, and beach bums. The term *cover* implies that we assign every vertex and face in the mesh to a specific group or region. Segmentation leaves no vertex or face unlabeled, so to speak. In this dissertation, we are interested in a region segmentation that partitions a mesh into contiguous regions that represent visual parts, as defined by the minima rule. The segmentation goal is essentially a change of representation that organizes a mesh in a higher-level description that is either more meaningful or more efficient—or both—for further analysis (Shapiro and Stockman, 2001). Recall our discussion of meaningful in Sec. 1.1 and thus our choice of the minima rule.

From our literature review in Sec. 2.3, three papers represent the state of the art in mesh segmentation (Vincent and Soille, 1991; Wu and Levine, 1997; Mangan and Whitaker, 1999). In our review, we noted the drawbacks to (Wu and Levine, 1997) and (Mangan and Whitaker, 1999) and suggested that (Vincent and Soille, 1991) avoids most of these drawbacks. Vincent and Soille propose a bottom-up flooding algorithm that we have identified as a robust algorithm for the minima rule. Unfortunately, the actual implementation of Vincent and Soille does not allow direct application of the minima rule. So, we seek a new flooding algorithm that is appropriate for our definition of height. As we will see, our definition of height is a directional definition that is necessary for proper implementation of the minima rule.

In this chapter, we describe a novel algorithm called *Fast Marching Watersheds* that implements watershed flooding and as such extends the state of the art beyond (Vincent and Soille, 1991) and more recent implementations (Rettmann et al., 2000; Rettmann et al., 2002). Specifically, we highlight the contributions of our algorithm as follows:

- creation of a fast and robust hill climbing algorithm for watershed segmentation on a triangle mesh data structure

- definition of a directional height map appropriate for the minima rule using local principal curvatures,

- application of morphological operations to improve the initial marker set for the above algorithm, and

- a fast implementation of connected component analysis on a triangle mesh to aid segmentation.

We outline the chapter as follows. We begin in Sec. 5.1 with a brief description of the watershed analogy to clarify the algorithm framework. Then, we present an overview of the algorithm, including a block diagram in Sec. 5.2. Next, Sec. 5.3 discusses the steps for generation of a marker set to initialize the algorithm, and Sec. 5.4 describes the connected components algorithm that distinctly labels the marker set. Sec. 5.5 presents the details of the hill climbing algorithm that grows the marker set into the final segmentation that covers the mesh. Then, we present in Sec. 5.6 the major contribution of this chapter, the definition of height for a minima rule segmentation. The final section is Sec. 5.7, which concludes with a few comments and remarks.

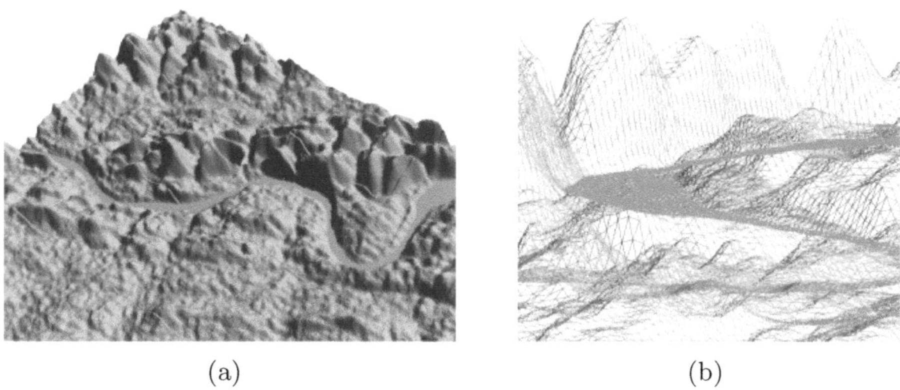

(a)   (b)

Figure 5.1: This triangle mesh helps us to visualize the watershed analogy. (a) Near Knoxville, the Tennessee River winds through the hills of this computer rendering of a mesh terrain. (b) A zoom view of the underlying mesh from (a).

5.1 Watershed Analogy

Suppose we have some triangle mesh that represents a surface model of a terrain such as the hills around the Tennessee River in Fig. 5.1. For top-down bobsledding, consider a drop of water placed at any point on the terrain. By gravity, the water will fall along the slope of the terrain to a valley, or in other words, the drop will slide like a bobsled to a local minimum. The beginning point where the water was first dropped and each of the points along its path to the valley belong to that valley's *watershed*, or *catchment basin*. Distinct valleys form distinct watershed regions, *i.e.* the segmentation regions. Although this description is a nice illustration, it leads to implementation problems such as temporary storage to track the path of the water drop, special procedures to handle flat plateaus, and output filtering to account for oversegmentation. Subsequently, bobsledding algorithms are typically more complex than the second formulation in the next paragraph.

A bottom-up flooding approach avoids these implementation problems. We can conceptually describe flooding as in (Baccar et al., 1996). For simplicity, consider the one-dimensional case in Fig. 5.2. We first punch holes in each valley of the terrain as in Fig. 5.2(b). We then begin flooding the terrain as in Fig. 5.2(c) from below by letting the water rise through the holes at a uniform rate. See Fig. 5.2(d). When the rising water in distinct catchment basins is about to merge, we build a dam to prevent the merging as in Fig. 5.2(e). The flooding will eventually climb beyond each terrain peak and only the tops of the dams will be visible above the water as in Fig. 5.2(f). The dam boundaries divide each watershed region and thus define our desired watershed segmentation. Since we grow outward from each valley, we do not need to temporarily track the water

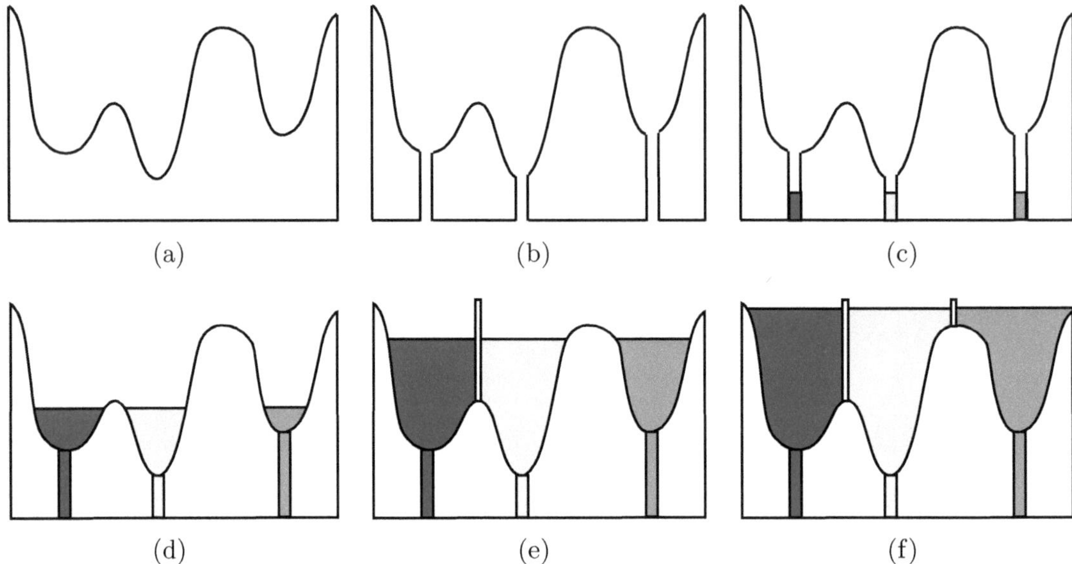

Figure 5.2: These simple one-dimensional terrains illustrate the flooding analogy of the watershed algorithm. (a) The empty terrain that we wish to segment. (b) We punch holes in the valleys of the terrain. (c) Flooding through the holes begins from below. (d) The flooding continues at a uniform rate. (e) When flood waters from distinct valleys meet, they do not merge, but rather we build a dam between them. (f) The process continues until each segment of the terrain is associated with a specific valley. For the case shown, we have three valleys.

flow until reaching a valley. We inherently know from which valley a particular water flow began. Also for plateaus, we assume the water creeps from the edge of the plateau until meeting another water flow or sliding down the other edge of the plateau. We thus avoid the ambiguity of deciding where a drop of water should flow on a plateau, as in bobsledding. Finally, we handle oversegmentation with a proper initialization procedure as discussed below.

We now return to Fig. 5.1. If we follow this same procedure for this terrain, we will have an overwhelming and useless segmentation since the terrain contains a very large number of valleys. The resulting segmentation would be grossly oversegmented as in bobsledding. The concept of a initial *marker set* helps alleviate this problem. To create a marker set, we threshold the terrain to a certain elevation, and the connected segments below that elevation become our initial valleys, *i.e.* our marker set. The examples in Fig. 2.8 best illustrate marker sets. In this figure, we see how different marker sets lead to different segmentations. Importantly, the final number of regions after segmentation is equal to the initial number of regions in the marker set. So, the key to a useful segmentation is the proper selection of a threshold to create the appropriate marker set. We will investigate the selection of a threshold in Sec. 5.3.1. A good marker set minimizes oversegmentation. Also, a good marker set reduces computational time as an initial threshold often segments 70% to 90% of the mesh. Thus, the watershed algorithm need only operate over the remaining 30% to 10% of the unsegmented mesh. This computational savings is an advantage of a marker set.

The question that we now ask is how does watershed flooding work when we no longer have a mesh that is a terrain map. Suppose our mesh is the now familiar mug example in Fig. 1.1. The mug does not have peaks and valleys like the terrain mesh. In fact, the topology of the mug, whose handle makes it a genus one surface, is very different from the genus zero topology of the terrain map. The answer is that we must formulate our mesh into a height map. For the terrain data, this formulation is straightforward and obvious. We conceptualize the triangle mesh as lying in the xy-plane and the z-coordinate serves as the height, which defines the peaks and valleys for our watersheds. For a mesh with arbitrary topology such as the mug that freely twists and turns in 3D, we can not simply select one coordinate as the "height". We must be a little more imaginative in our formulation. A solution is to introduce a fourth dimension to each vertex of the mesh. This additional dimension is a value that serves as the height of the mesh at that vertex. What is this new fourth value? Well, the answer is dependent on our segmentation application. For this dissertation the height value is the local curvature at a particular vertex. For some other application, it might be the color of the surface. A variety of different height map definitions are possible. With a height value at each vertex, a watershed algorithm for a mesh treats the surface, regardless of topology, as a planar surface as in the terrain case. In other words, a watershed algorithm considers surfaces with arbitrary topology to locally appear planar with the additional height values defining peaks and valleys. With this concept, we next define our watershed algorithm.

5.2 Algorithm Overview

Our Fast Marching Watersheds—both the name and the algorithm—derive inspiration from Fast Marching Methods (Kimmel and Sethian, 1998) for computing shortest geodesic paths on a mesh. Kimmel and Sethian's algorithm employs a heap data structure to control the geodesic "walk" across the surface of the mesh. For our watershed algorithm, we use a similar heap structure to control the "flooding" across the vertices of the mesh. Unlike Kimmel and Sethian, however, our heap keys on local height values, and not cumulative geodesic distances, and it tracks the progression of regions, and not geodesic paths. These differences mark

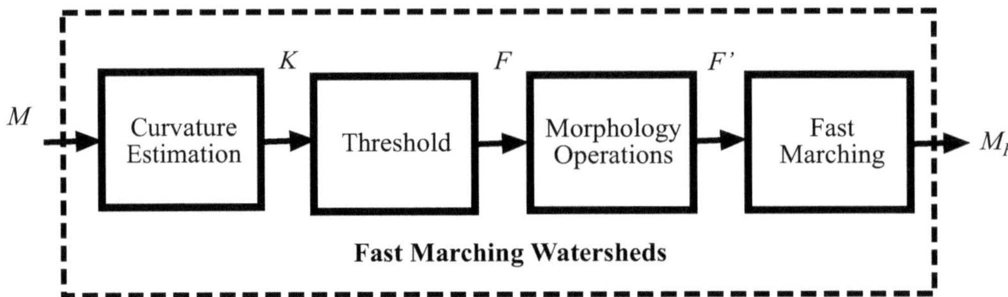

Figure 5.3: Block diagram of the Fast Marching Watersheds algorithm. The input is a triangle mesh, and the output is the set of watershed regions.

our contribution of extending Kimmel and Sethian's algorithm into the wholly different application of mesh segmentation and the formulation of a flooding-based algorithm. A block diagram of the integrated algorithm appears in Fig. 5.3.

The input to the algorithm as shown in the figure is a triangle mesh M and height map H for the vertices of the mesh. We specify the height map as the following set

$$H = \{h_0, \ldots, h_{n-1}\} \tag{5.1}$$

where n is the number of vertices in the mesh and h_i is the "height" at vertex i. We define the actual value of h_i for our minima rule algorithm in a later section. The first two blocks in the diagram create the marker set F' to initialize the watershed algorithm. Then, the third block uniquely labels the regions $M_r' \subset M_R'$ individually through a connected component analysis where r is the specific label and R is the number of such labels. Finally, the last block grows the M_r' marker regions into the final watershed catchment basins $M_r \subset M_R$ that covers the entire mesh.

$$M = \bigcup_{r=0}^{R-1} M_r = M_R \tag{5.2}$$

such that $M_r \cap M_s = \emptyset$ when $r \neq s$. Note that $M = M_R$ where M_R simply denotes the segmented version of M into R regions. Fig. 5.4 shows two examples for each step. We now look in depth at each of the blocks in the following sections.

5.3 Marker Set

The first two blocks in Fig. 5.3 establish the marker set for the initialization of the watershed algorithm. These blocks label the mesh in a binary fashion such that each vertex of the mesh either belongs to the marker set or does not. The initial step is a straightforward thresholding of the height map values in (5.1). Then, the morphology operations in the second step clean up the threshold regions and

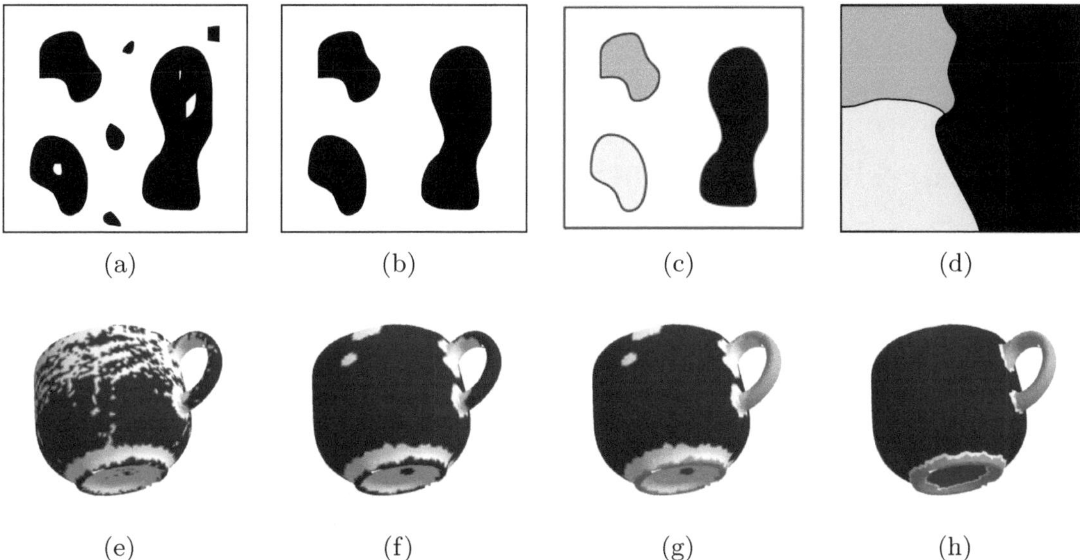

(a) (b) (c) (d)

(e) (f) (g) (h)

Figure 5.4: These simple examples show the progression of the steps in the Fast Marching Watersheds algorithm. (a) A 2D planar example with the initial threshold yielding black blobs in a binary representation. (b) Morphology operators close the blobs and remove unwanted smaller regions to establish a clean marker set. (c) Connected components uniquely labels each blob, in this case with a specific grey scale. (d) At conclusion of marching, the whole plane is segmented into the final regions. Since we began with three markers in (c), we end with three regions in (d). (e–h) Same steps as previous example except for the mug.

Table 5.1: Two definitions of different foregrounds for a single threshold t.

Threshold Above	Threshold Below
$h_i > t$	$h_i < t$

thereby establish a robust marker set. By robust, we mean to say a marker set that avoids significant oversegmentation at the completion of the watershed algorithm.

5.3.1 Threshold Segmentation

Thresholding of gray-scale images is well known (Shapiro and Stockman, 2001), and thresholding of a height map, such as H above, over a mesh is not much different. Thresholding yields a binary map of the vertices where some vertices are—in image processing terminology—*foreground* and the rest are *background*. So, thresholding is a group segmentation since we do not care about the topological relationship of the vertices. In the simplest case, we choose a single threshold value t and designate whether the foreground is *threshold above* or *threshold below*. In other words, for threshold above if $h_i > t$, then vertex i is foreground. Conversely, for threshold below, if $h_i < t$, then vertex i is foreground. Table 5.1 summarizes these definitions. Thresholding above yields the foreground set

$$F = \{\, v_i \mid h_i > t \,\} \tag{5.3}$$

where $i \in \{0, \ldots, n-1\}$ and thus $v_i \in V$ denotes a specific vertex in the mesh $M = (K, V)$. Recall the mesh topology is K and the geometry is the vertex set V. We have $F \subset V$ and therefore a background set F_B.

$$F_B = V \setminus F \tag{5.4}$$

where the operator $\setminus$ denotes set difference. Obviously, the crux of thresholding is choosing t where (Gonzalez and Woods, 1993; Shapiro and Stockman, 2001) offer a variety of strategies.

Fortunately, for our application, the minima rule defines a clear choice for the threshold, and thus complex strategies for optimal thresholds are unnecessary. With the minima rule, part boundaries occur at negative minima of curvature. Thus, $t = 0$ is a natural choice. If we threshold above $t = 0$, we can create a marker set for the parts of an object and flood the watersheds until reaching the

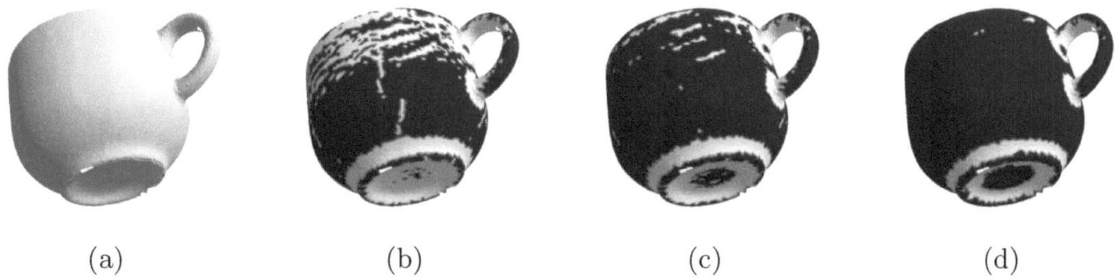

(a)   (b)   (c)   (d)

Figure 5.5: An offset from zero improves the threshold segmenation. (a) Original mesh without thresholding. (b) A binary map on the mug where black indicates foreground vertices F with positive curvature and white indicates background ones F_B with negative curvature. The threshold is set at $t = 0$. (c) Another binary map where we have offset the threshold to a slightly more negative value. Observe the reduction in the size and number of white patches. (d) A third threshold example that is even more negative but still close to zero in a relative sense.

negative minima boundaries. Consider Fig. 5.5. In Fig. 5.5(a), we again show the surface mesh for the mug. If we apply the $t = 0$ threshold on curvature values, we create the threshold-above map in Fig. 5.5(b). With an understanding of curvature, we may not necessarily have expected this result. The small white patches on the cup of the mug are at areas where we probably expect strictly positive curvature values. Ideally, we should only have white background around the handle joints and bottom cusp. The problem is that slight errors in curvature estimation have led to negative values for these small areas.* A practical solution to minor estimation errors, such as these errors, is to offset the threshold from zero to some moderately negative value just below zero as in Figs. 5.5(c) and 5.5(d). If we compare these latter figures to the previous one, we notice a reduction in the size and quantity of erroneous white patches. We do, however, preserve as white background—and thus not part of the marker set—the negative curvatures around the handle and base as we desire. With a simple offset of the threshold from zero, we gain improvement in the marker set. We can further improve this initial set using morphological operations.

---

*Although we are using Normal Vector Voting from Ch. 4 for our curvature estimation, we have purposefully set the voting neighborhood of that algorithm to a small value to introduce some inaccuracies in the curvature measures. Our intent is to investigate ways to overcome such problems through our watershed algorithm and thus to build robustness into the segmentation process itself.

5.3.2 Morphological Operations

Thresholding is usually only successful under highly controlled conditions. As we found in Ch. 4, curvature estimation is definitely not a controlled environment. In fact, the moderate variations in thresholds for Fig. 5.5 demonstrate the sensitivity of thresholding to curvature. To overcome this sensitivity, we suggest *mathematical morphology* as a post-processing tool to improve the marker set. To clarify terms and notation, we briefly review the basic operations of mathematical morphology, and refer the reader to (Gonzalez and Woods, 2002) for additional details. Specifically, we discuss the two fundamental operations of *dilation* and *erosion*, which by definition lead to the more powerful operations of *opening* and *closing*.

The language of morphology is set theory. Consider two sets $A, B \subset \mathbb{R}^2$.[†] We define dilation as

$$A \oplus B = \{p \mid [(\hat{B})_p \cap A] \subseteq A\} \tag{5.5}$$

where $\hat{B}$ denotes the reflection of B while the subscript p denotes shifting this reflection by $p \subset \mathbb{R}^2$. We call set B the *structuring element*, and typically B is a symmetric set such that $B = \hat{B}$. Similarly, we can define erosion in the following equation

$$A \ominus B = \{p \mid (\hat{B})_p \subseteq A\} . \tag{5.6}$$

Dilation expands A with the structure of B while erosion shrinks it. With these definitions, we can build the definitions for opening and closing. An opening is simply an erosion followed by a dilation.

$$A \circ B = (A \ominus B) \oplus B . \tag{5.7}$$

Opening generally smoothes the contour of a shape, breaks narrow isthmuses, and eliminates thin protrusions. A closing is just the opposite sequence.

$$A \bullet B = (A \oplus B) \ominus B . \tag{5.8}$$

Closing tends to smooth the contour of a shape as well but, in contrast to opening, it generally fuses narrow breaks and long thin gulfs, eliminates small holes, and fills gaps in the contour. The power of opening and closing is that used in conjunction with each other, they clean the shape boundaries and interiors. Fig. 5.6(c) is a much better marker set than Fig. 5.6(a).

[†]These sets are in 2D to simplify the definition of the morphology operators. Sets that represent two-manifold surfaces embedded in 3D require more complicated discussion that adds little to the understanding of the operators. So, for now, we simply use these 2D definitions.

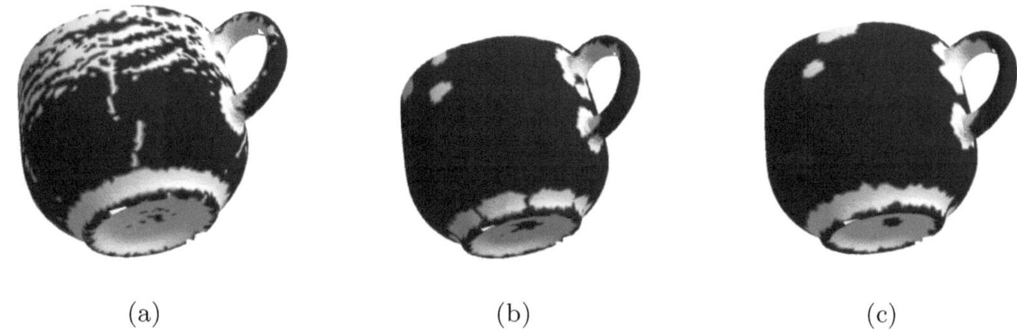

(a)   (b)   (c)

Figure 5.6: Morphology operations improve the marker set even when the initial threshold may not be perfect. These figures show the progression of a common combination of morphological operations, which is a closing followed by and opening. (a) Initial curvature threshold with $t = 0$ for the mug. (b) Closing by applying dilation followed by erosion to (a). (c) Opening by applying erosion followed by dilation to (b).

Rössl et al. (Rössl et al., 2000) discuss the implementation of Eqs. (5.5)–(5.8) for the arbitrary connectivity of a triangle mesh, embedded in 3D. Since we claim no contributions with regard to this implementation, we avoid the details of that discussion. We instead emphasize two prominent features of their implementation that one must consider. First, as (Ronse, 1989) and (Roerdink, 1996) note, "mathematical morphology is flat", which is to say, it is unclear how to define morphology on curved smooth surfaces. Roerdink (Roerdink, 1994) takes a first attempt towards morphological analysis on surfaces using techniques from differential geometry such as geodesic paths, parallel transport and exponential maps. The difficulty in extending these concepts to triangle meshes has led Rössl et al. to redefine the morphological operators in a limited manner with only a disk-like structuring element. They base their structuring element on the k-ring neighborhood about a vertex where the k ring defines the "radius" of the disk element. This definition is a topological one and not a geodesic one as (Roerdink, 1994) seeks. Heijmans (Heijmans, 1994) refers to such approaches as *graph morphology*. The second issue regards implementation of erosion and dilation beyond a one-ring disk to a k-ring disk. Rössl et al. state that iterative application of successive one-ring dilations and erosions is equivalent to a simultaneous k-ring operation. The former approach has some computational advantages over the latter, and consequently, we use a similar approach for our implementation.

So, from Rössl et al. (Rössl et al., 2000), we apply an opening and a closing operation in sequence to clean our initial threshold set F. The user must specify a single parameter k, the size of the disk structuring element. We choose $k = 1$ for most applications. We outline our morphology operations on F to improve the marker set F' as follows:

$$F' = (F \bullet D_k) \circ D_k \tag{5.9}$$

where D_k specifies the special disk element of Rössl et al. An example disk element is D_1, which consists of the vertex v_i along with each v_j in the one-ring neighborhood of v_i. Again, consider Fig. 5.6. For this figure, the initial threshold is exactly zero and thus yields the poor initial markers in Fig. 5.6(a). We see in the sequence of figures how a closing with D_1, Fig 5.6(b), and then an opening with D_1, Fig. 5.6(c), improves the marker set considerably. The closing bridges the small white gaps around the cup while the opening eliminates the small isthmuses that appear around the base as a result of closing. The next step is to uniquely label this marker set, which we address in the next section with connected component analysis.

5.4 Connected Components

As with morphology, connected component analysis is well-known in image processing (Gonzalez and Woods, 2002; Shapiro and Stockman, 2001), and Figs 5.4(b) and 5.4(c) illustrate the idea with a simple example. The basic concept is to generate unique labels for each connected region in the binary set F' from (5.9) and to identify the connectedness of each region. Two vertices $v_i, v_j \in F'$ are *connected* if there exists a path along mesh edges between them consisting entirely of vertices in F'. In other words, as we walk along the edges of the mesh from v_i to v_j we encounter vertices v_o. If each $v_o \in F'$, then v_i is indeed connected to v_j. A *connected component* M_r' is the set of such vertices that are connected to each other. Thus, we have the following relationship:

$$F' = \bigcup_{r=0}^{R-1} M_r' \tag{5.10}$$

where M_r' is a connected region within F' and R is the total number of such regions. The M_r' regions are disjoint and have no overlap such that

$$M_r' \cap M_s' = \emptyset \ \forall \, r \neq s \,. \tag{5.11}$$

Note that we grow M_r' marker into the final segmentation region M_r. We present an algorithm to find the connected components on a triangle mesh in Alg. 5.7.

We use a boolean array `visited` to track vertices that we have labeled. We initialize `visited` for $v \notin F'$ as `true` and for $v \in F'$ as `false`, such that we avoid

input : A triangle surface mesh M and a marker set F'.
output: An array label whose elements are associated with vertices in M. The labels are from the set $\{-1, 0, \ldots, (R-1)\}$ where -1 denotes "unlabeled".

initialize array label *as "unlabeled"*;
initialize boolean array visited *with* F';
$r \leftarrow 0$;
foreach v *in* M **do**
    if *not* visited $[v]$ **then**
        `Clear`(frontier);
        `Push`(frontier, v);
        repeat
            $v_i \leftarrow$ `Pop`(frontier);
            label $[v_i] \leftarrow r$;
            visited $[v_i] \leftarrow$ **true**;
            foreach v_j *in umbrella neighborhood of* v_i **do**
                if *not* visited $[v_j]$ **then** `Push`(frontier, v_j) ;
                ;
            end
        until `Empty`(frontier);
    end
    $++r$;
end

Figure 5.7: Connected Components Algorithm for a surface mesh data structure.

vertices not in the marker set. The queue `frontier` tracks the boundary of a region as the labeling extends to connected foreground vertices. Note that the inner **foreach** loop circulates around the umbrella neighborhood of each vertex. This loop requires a mesh data structure in M capable of maintaining incidence information of vertices and faces (Kettner, 1999). The algorithm complexity is linear $O(n)$ where n is the number of vertices for M. This algorithm is not necessarily a contribution to the state of the art, but we have not found a connected component algorithm explicitly delineated in the literature for a triangle mesh. With this algorithm, we now have our marker set uniquely labeled, and thus we can proceed with the watershed algorithm.

5.5   Watershed Algorithm

To this point, we have specified a marker set and labeled that set into distinct catchment basins. We now seek an algorithm to grow these catchment basins to segment the entire mesh. The selection of a particular watershed algorithm is not straightforward and often confusing since the image processing literature lacks thoughtful distinctions between algorithm specification and implementation (Roerdink and Meijster, 2001). Our interest in implementing a mesh algorithm, and not an image one, further complicates matters. Roerdink and Meijster, however, attempt to bring some order—at least for image processing—to the situation through a critical review of several watershed definitions and the associated algorithms that follow from those definitions. From this review, we have selected the hill climbing algorithm. Our motivation for choosing this algorithm is that it is the simplest since we do not have to compute geodesic distances and that the reliance on a local height computation allows us to implement the minima rule. Our implementation of this algorithm appears in Alg. 5.8. This algorithm begins with the marker set F' and grows that set until we have segmented the whole mesh. The background becomes empty, $F'_B = \emptyset$. The close-up views in Fig. 5.10 illustrate this sequence.

Although both algorithms are flooding methods, our hill climbing algorithm differs slightly from the immersion algorithm of (Vincent and Soille, 1991) as Fig. 5.11 demonstrates. The main advantage as (Roerdink and Meijster, 2001) notes is the algorithm simplicity since it does not require merging operations as flooding progresses. The heart of our implementation is the `heap` data structure that controls the hill climbing process and the procedure `ExtendBoundary()` that populates this heap. The key into the heap is local curvature such that positive curvature keys bubble to the top and more negative ones sink to the bottom. The `PopHeap()` function pulls from the top of the heap to begin each stage of the marching process. The data associated with a key specifies which vertex to march

input : A triangle surface mesh M and an array label with some
"unlabeled" elements.
output: The array label contains no elements that are "unlabeled."

initialize heap;
foreach v_i *in* M **do**
    if label $[v_i]$ **then** ExtendBoundary(v_i, heap, label) ;
    ;
end
while *not* Empty(heap) **do**
    data $\leftarrow$ PopHeap(heap);
    $v_i \leftarrow$ ExtractVertex(data);
    if *not* label $[v_i]$ **then**
        label $[v_i] \leftarrow$ ExtractLabel(data);
        ExtendBoundary(v_i, heap, label);
    end
end

Figure 5.8: Fast Marching Watershed Algorithm for a surface mesh data structure. The power of this algorithm is in the **heap** data structure and in the procedure ExtendBoundary(). The **heap** basically tells the algorithm where to march next while ExtendBoundary() finds potential candidates for places to march.

input : A mesh vertex v_i and an array label and heap
output: Marches and extends heap with umbrella neighbors of v_i.

initialize data;
foreach v_j *in umbrella neighborhood of* v_i **do**
    if *not* label $[v_j]$ **then**
        data $\leftarrow$ CreateHeapData(v_j, label $[v_i]$);
        key $\leftarrow$ ComputeDirectionalHeight(v_i, v_j);
        PushHeap(heap, key, data);
    end
end

Figure 5.9: ExtendBoundary(vertex, heap, label) is a procedure to extend the marching boundary in the Fast Marching Watersheds algorithm. The ComputeDirectionalHeight() function computes the directional curvature, relative to the minima rule.

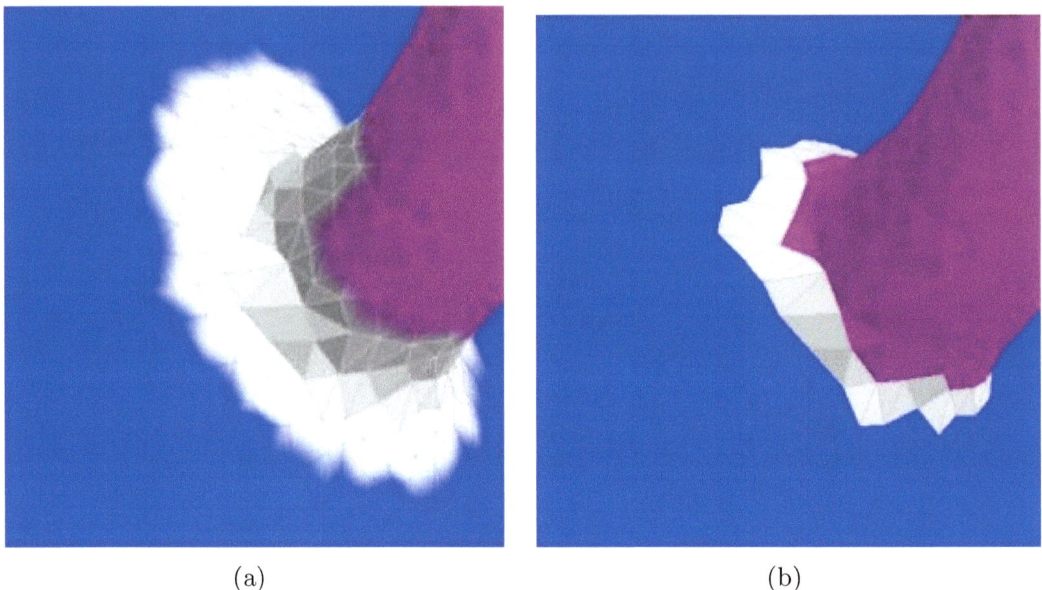

(a)   (b)

Figure 5.10: These close-up views show the progression from the marker set (a) to the final segmentation (b). The white triangles are unsegmented in (a). In (b), these triangles represent the dams that divide the catchments basins for the cup and the handle of the mug.

to and to what watershed it potentially belongs. If we pop a vertex from the heap and it is already labeled, we simply pop the next one. This climbing heap is similar to the marching heap found in Fast Marching Methods (Kimmel and Sethian, 1998). In fact, our implementations of Fast Marching Methods for Ch. 4 and Fast Marching Watersheds in Alg. 5.8 are almost identical. As with Fast Marching Methods, Fast Marching Watersheds have computational complexity $O(n \log n)$ where n is the number of vertices in the mesh since the `PushHeap()` operation is an $O(\log n)$ operation. This order assumes that we must segment each of the n vertices, but as we have noted, with an initial marker set we typically have anywhere from 70–90% of the vertices already segmented. So, n is usually much less than the actual number of vertices in the entire mesh. The primary difference however between Fast Marching Methods for geodesics and our Fast Marching Watersheds for segmentation is the function `ComputeDirectionalHeight()`. This function is the most significant contribution of this chapter and is the impetus of the minima rule decomposition. As such, we outline the equations that govern this function in the next section.

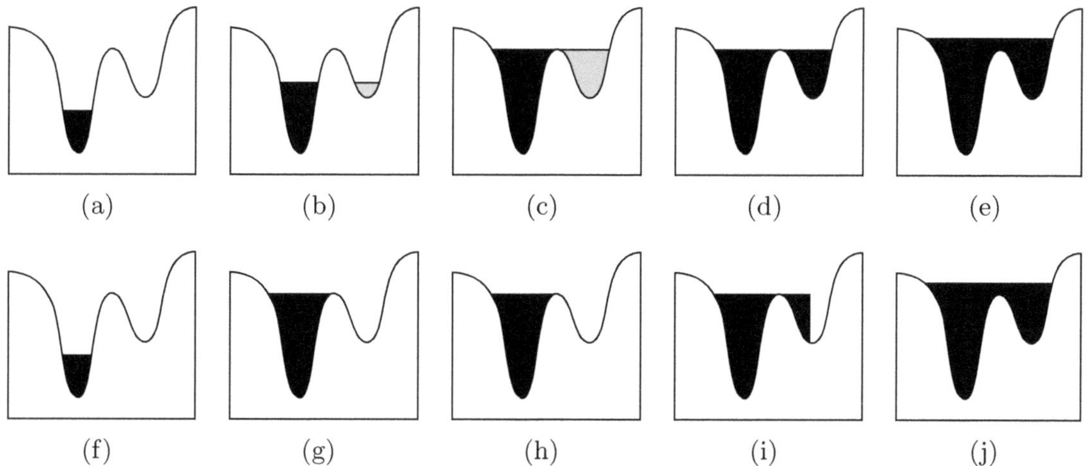

Figure 5.11: Comparison of two flooding methods. The top row (a–e) shows the immersion approach (Vincent and Soille, 1991) and the bottom row (f–j) shows the hill climbing approach of our algorithm. Note the merging between (c) and (d) for immersion compared to hill climbing in (h) and (i). Both sequences are a trivial case where we begin with a single marker in (a) and (f) and end with a single catchment basin in (e) and (j), respectively.

5.6   Directional Height

In Sec. 5.2, we presented a set definition of water height for each mesh vertex. Recall Eq. (5.1). With this equation, each vertex v_i has a corresponding height h_i. This global definition of height is common in watershed algorithms, yet for the minima rule, we need a more local definition. We need a definition such that we specify the height h_{ji} from vertex v_j to vertex v_i, relative to the local curvature between these vertices. Eq. (2.1) serves as a starting point. This equation specifies the normal curvature along a particular direction. For our local curvature height, this direction is from v_j to v_i. The tangent unit vector T_{ji} at v_j associated with this direction is as follows:

$$T_{ji} = \frac{\mathbf{T}_j(v_i - v_j)}{\|v_i - v_j\|} \tag{5.12}$$

where $\mathbf{T}_j$ is a 3×3 matrix that projects the direction vector $(v_i - v_j)$ into the tangent plane of v_j. We define this matrix as follows:

$$\mathbf{T}_j = \mathbf{I} - N_j N_j^t \tag{5.13}$$

where $\mathbf{I}$ is the identity matrix and the vector N_j is the surface normal at v_j. The superscript t denotes the transpose and thus leads to an outer product operation.

The angle θ in Eq. 2.1 is the angle between T_{ji} and the maximum principal direction T_1 at vertex v_j. The dot product of these two vectors yields the cosine of this angle.

$$\cos \theta_i = T_{ji}^t T_1 \ . \tag{5.14}$$

With the above equations, we can now compute the normal curvature κ_{ji}.

$$\kappa_{ji} = \kappa_j^1 \cos^2(\theta_i) + \kappa_j^2 \sin^2(\theta_i) \tag{5.15}$$

where κ_j^1 and κ_j^2 are the maximum and minimum principal curvatures at v_j, respectively. Thus, the `ComputeDirectionalHeight()` function in the `ExtendBoundary()` procedure returns κ_{ji} as the directional height from v_j to v_i. This definition of height allows the watershed algorithm to flow along vertices with similar values of curvature but impedes climbing up negative curvature hills. This definition of height is interesting as it implies that the height h_{ji} for a vertex v_j is dependent on the direction that we approach the vertex.

$$h_{ji} = \kappa_{ji} \ . \tag{5.16}$$

This formulation is our major contribution to the watershed literature, especially in the context of minima rule decomposition and mesh segmentation.

5.7   Remarks

In this chapter, the Fast Marching Watershed algorithm leads to a minima rule decomposition of triangle meshes. See Fig. 5.4. At this point, we have not been rigorous in our discussion of the figurative water height for this algorithm. In our presentation, we have toggled between global height h_i and local directional height h_{ji}. We acknowledge our lack of distinction and attempt to clarify now. When we threshold, we do consider the water height as global. We set the height equal to the minimum principal curvature at each vertex, $h_i = \kappa_i^2$. However, when we segment after thresholding, we consider the water height as directional. We set the height h_{ji} as in Eq. 5.15. So, we actually use both definitions of height depending on the particular stage of the algorithm.

Another issue is the choice for the threshold value t from Sec. 5.3.1. As noted in that section, the minima rule suggests that $t = 0$ is the natural choice, but implementation results suggest that a slight negative offset from zero improves results. The question is how far to offset. We have defined this offset as a user parameter for our algorithm. Consider the set of vertices V^- with negative minimum

Table 5.2: Parameters for Fast Marching Watershed Algorithm.

Parameter	Range	Equation	Typical Value	Comments
k, D_k	$k \in \mathbb{Z}^+$	(5.9)	$k = 1$	Determines the size of the disk structuring element for the morphological operators. Since k denotes a k-ring, we choose k relative to the feature dimensions of the surface mesh.
α	$0 \le \alpha \le 1$	(5.18)	$\alpha = 0.3$	Determines the threshold t for the marker set. We specify the percentage α instead of t directly to avoid scaling issues.

principal curvatures such that

$$V^- = \{\, v_i \mid \kappa_i^2 \le 0 \,\} \tag{5.17}$$

where κ_i^2 is the minimum principal curvature for vertex v_i. We can average across this set and establish our threshold as a percentage α of this average as follows

$$t = \frac{\alpha}{n^-} \sum_{v_i \in V^-} \kappa_i^2 \tag{5.18}$$

where n^- is the number of vertices in V^-. Thus, the user chooses α instead of the threshold t directly. Our implementation is not sensitive to the choice of α, but we suggest that $\alpha = 0.3$ yields good results for most applications.

With the above parameter, the Fast Marching Watersheds algorithm has a total of two user parameters that control the segmentation process. The other parameter is the size k of the disk structuring D_k from Sec. 5.3.2. We summarize both of these parameters in Table 5.2. The objective of these parameters is to limit oversegmentation, but as with most watershed systems, oversegmentation is inevitable. In the next chapter, we address this problem with a metric to measure the quality of the segmentation output in terms of the visual salience of parts.

Chapter 6

Shape Measure: Part Saliency Metric

A shortcoming of watershed segmentation in image processing is oversegmentation (Vincent and Soille, 1991; Roerdink and Meijster, 2001; Gonzalez and Woods, 2002) and is equally true in mesh segmentation as well (Mangan and Whitaker, 1999; Rettmann et al., 2000; Rettmann et al., 2002). We have attempted to build robustness into our system through our curvature estimation with Normal Vector Voting and our marker selection with Fast Marching Watersheds to overcome or, at least to a certain degree, minimize this problem. We have no guarantees however that our methods eliminate the problem entirely. We have no way of knowing the quality of our segmentation. We therefore propose the *Part Saliency Metric*. This measure is the salience, i.e. the significance, each segmented part contributes to the entire mesh, and it quantifies the importance of a part relative to the whole mesh, which in essence evaluates the quality of a part as a part itself. Thus, the metric allows a ranking of the parts from least to greatest in terms of this quality. With such a ranking, we can address oversegmentation by merging the least salient parts with more salient ones until we eliminate enough "bad" parts to reach a "good" segmentation. As we will see, the terms *bad* and *good* are debatable and open for discussion. For now, we delay such a discussion, but we note that the Part Saliency Metric serves as a quantitative measure to begin such a discussion and constrains the debate to selection of a threshold to demarcate bad and good.

As an illustration, consider once again the now infamous mug in Figs. 6.1 and continued in 6.2. This sequence shows a potential progression of part merges based on the Part Saliency Metric. The initial segmentation in Fig. 6.1(a) reflects five parts, which most observers would agree is too many parts. The knob at the top junction of the handle with the cup and the flat bottom underneath are

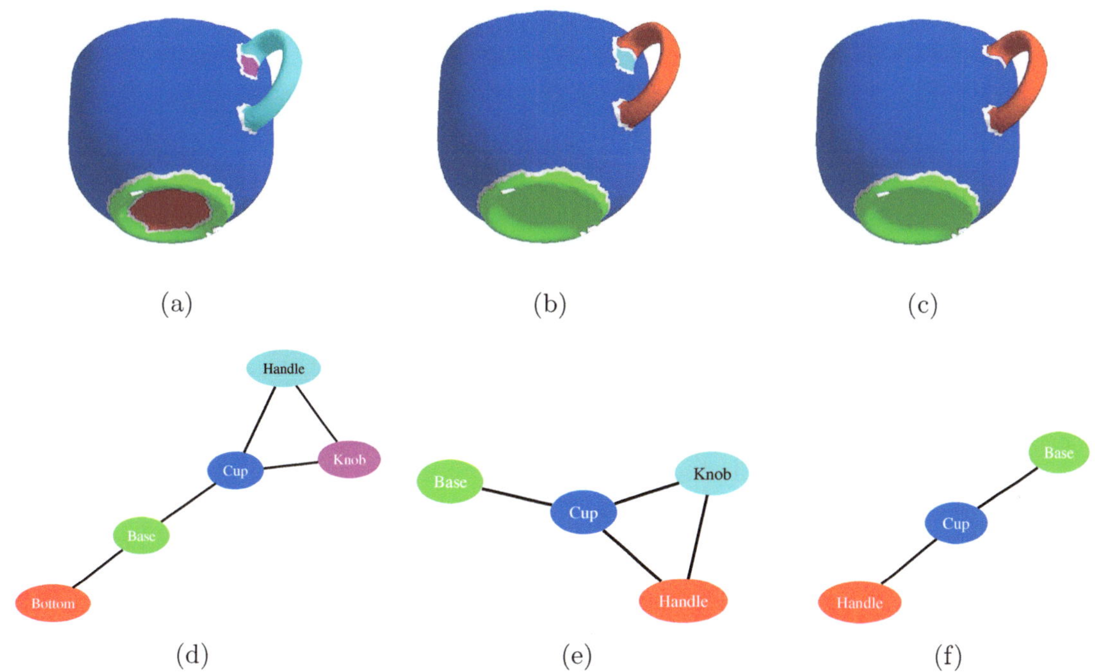

Figure 6.1: Sequence of region merges where at each stage we combine the least salient part with a more salient one. The top row shows a color labeling of the mug for the progression of part merges while the bottom row shows corresponding Part Adjacency Graphs for these segmentations. (a,d) Five parts: cup, handle, base, bottom, and knob. (b,e) Four parts: bottom merges with base. (c,f) Three parts: knob merges with handle.

oversegmentations. With the methods in this chapter, we can compute the salience of the knob and the bottom, and we thus discover that they are indeed the least important parts of the segmentation. So, we can trim them from the segmentation and merge with the handle as in Fig. 6.1(b) and 6.1(c). We can continue to merge the base with the cup in Fig. 6.2(b) and the handle with the cup too in Fig. 6.2(c). The Part Saliency Metric governs the decision as to which part is the next one to merge, and by design, this process agrees with our visual perception of the parts. Most observers would rank the parts from least to most salient as the bottom, the knob, the base, the handle, and finally the cup, which the sequence we have just outlined follows. This result is because we have designed our metric in accordance with the human vision theory in (Hoffman and Singh, 1997).

Another way to look at this progression is in the opposite direction. A human when first viewing the mug in Fig. 6.2(c) most likely perceives the handle as the most significant part of the mug aside from the cup itself. Then, if pressed to identify another part of the mug, she would probably point out the base as

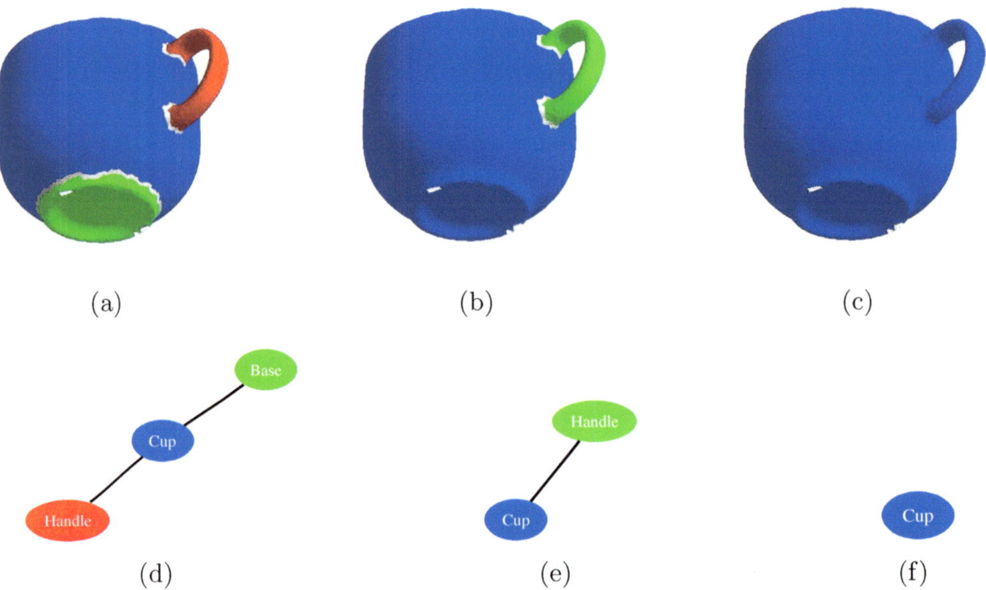

(a)   (b)   (c)

(d)   (e)   (f)

Figure 6.2: Continued from Fig. 6.1. Sequence of region merges where at each stage we combine the least salient part with a more salient one. The top row shows a color labeling of the mug for the progression of part merges while the bottom row shows corresponding Part Adjacency Graphs for these segmentations. (a,d) Three parts: base, handle, and cup. (b,e) Two parts: base merges with cup. (c,f) One part: handle merges with cup.

the third part of the mug. At this point, however, our human observer would probably stop and may never select the bottom or knob as distinct parts of the mug. She somehow thresholds her notion of parts and only identifies "good" parts. The minima rule captures her perception of parts, but the rule does not address the visual salience of the resulting parts. Our goal in this chapter is to seek an algorithm that allows us to measure and threshold the visual importance of a part and thus keep only good parts.

Leveraging the perceptual theory of the minima rule, Hoffman and Singh propose that the visual salience of a part depends on at least three factors: its relative size to the whole object, the degree that it protrudes from the object, and the strength of its boundaries. Recall Fig. 2.10. They support their theory with results from psychophysical experiments. In this chapter, we translate this human vision theory into a computer vision algorithm that computes the salience of parts from a segmentation of a triangle mesh. We apply this metric to the problem of oversegmentation through a filter and merge algorithm that trims the least salient parts. With the development of this Part Saliency Metric and the accompanying merge algorithm, we claim the following contributions to the state of the art:

- implementation of a metric for the visual salience of a minima rule part for triangle meshes,

- creation of an adjacency graph for representation of our minima rule decomposition, and

- application of the saliency metric and the graph as a filter and merge postprocessing step.

We outline the chapter as follows. In Sec. 6.1, we first discuss an overview of the algorithm to computer part salience and trim least salient parts. Then, in Sec. 6.2, we present our method for generating an adjacency graph to represent the part decomposition. This section also includes an interesting discussion of the four-color problem from graph theory as a means to improve the coloring of our segmentation labeling scheme. Next, Sec. 6.3 is the most important section of this chapter. We formulate the equations for the Part Saliency Metric within this section and demonstrate a methodology for computing the metric. In Sec. 6.4, we define an algorithm for merging the least salient parts of an object with other more salient ones. Finally, we conclude in Sec. 6.5 with a few closing remarks concerning the user parameters in this chapter.

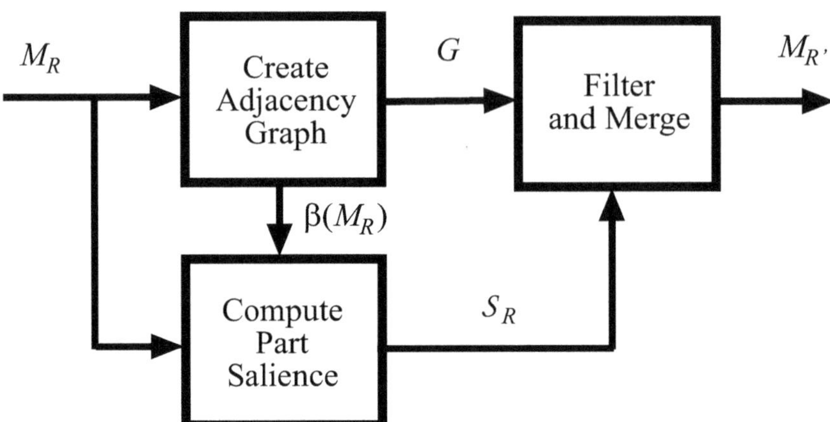

Figure 6.3: Block diagram of the filter and merge algorithm using the Part Saliency Metric. The input is the part regions $M_r \subset M_R$ of the triangle mesh M, and the output is a reduced set of those regions, $M_{r'} \subset M_{R'}$ where $R' < R$.

6.1   Algorithm Overview

The algorithm associated with the Part Saliency Metric first creates an adjacency graph of the segmented mesh, then computes the saliency measure for each part, and finally merges the least salient parts with other more salient ones. Fig. 6.3 illustrates this process as a block diagram. The adjacency-graph block takes the segmented mesh M_R and searches for neighboring parts to create an *undirected graph* G, which entails isolating the boundary vertices. We label the whole set of boundaries for the mesh as $\beta(M_R)$. The bottom block is the implementation of the Part Saliency Metric. The input to this block is the parts $M_r \subset M_R$ and the associated boundaries $\beta(M_r) \subset \beta(M_R)$, and the output is the set S_R of saliency measures for each part.

$$S_R = \{S_1, \ldots, S_{R-1}\} \, . \tag{6.1}$$

The final block uses this salience set S_R and the graph G to merge the least salient parts to form the final segmentation $M_{R'}$ with $R' < R$ total parts. We now discuss each block in more detail.

6.2   Part Adjacency Graph

The first step in this chapter is to create a *Part Adjacency Graph, $G = (R, E)$*, from the segmented mesh M_R. The undirected graph G consists of a set of part nodes R and a set of adjacency edges E. Each edge in E is an unordered pair (r, s) where $r, s \in R$. Example graphs appear in Fig. 6.1 and 6.2. These graphs show the adjacency relationship among the parts of a mug and are similar to the region adjacency graphs (Shapiro and Stockman, 2001) from image processing and the surface adjacency graphs (Hoover et al., 1998) from range processing. For the data structures to represent and manipulate G, we use the *Boost Graph Library* (Siek et al., 1999; Siek et al., 2001). This C++ library provides generic graph classes and algorithms. The translation of M_R into G mainly requires the identification of the region boundaries $\beta(M_R)$. We first outline an algorithm to find $\beta(M_R)$, and then we briefly detour to a side topic known as the four-color problem, which we use to color our segmentation graph.

6.2.1 Boundary Identification

Given a segmented mesh M_R, we may not necessarily know the vertices, $v \in \beta(M_R)$, that lie at region boundaries. Recall Fig. 5.10(b). In this zoom view, we see the shaded triangles for both the handle and the cup parts and the white triangles between them. These white triangles are the part boundaries. Although we know which vertices are at the boundary from this figure, the computer does not. We need a computer algorithm that finds the boundary vertices $\beta(M_r)$ for a part M_r. We present such an algorithm in Alg. 6.4. Note that this algorithm is similar to the connected components algorithm in the previous chapter.

This algorithm loops through each vertex v of the mesh. When the algorithm finds a boundary vertex, it traverses clockwise around that particular boundary and tags each vertex as "visited" thereby avoiding multiple walks around a boundary from the outer loop. The walking process requires two special functions to properly handle singularities such as in Fig. 6.5. In this figure, the problem is that the boundary passes twice through v and thus requires consideration of the topology of the boundary. The first function is `FindFirstClockwiseBoundary()`. This function takes as input a boundary vertex v and returns a halfedge h that is on the boundary and points to v. The concept of a halfedge is found in (Kettner, 1999), and we illustrate it in Fig. 6.6. Since each vertex on the boundary has at least two boundary halfedges that point to it, we specify h as a halfedge that we encounter if we start inside the region and walk clockwise around v until reaching the boundary. For most boundary vertices, this specification of h is unique, but for others, two or more halfedges may satisfy the specification. As an example, the singular vertex v in Fig. 6.5 has two such halfedges. The particular one that the function returns is arbitrary with respect to our application and not necessary to uniquely specify since we only seek a starting h for boundary traversal. Our walk will eventually return to the other boundary halfedges at v. The second function is `FindNextClockwiseBoundary()` and is the crux of the whole algorithm. This function takes as input a halfedge h that meets the previous specifications and returns the next clockwise halfedge on the boundary around v. To implement

```
input  : A segmented mesh M_R.
output : An array β whose R elements are linked lists containing the
         boundaries for each part.

initialize boolean array visited as false;
initialize array β with R null lists;
foreach v in M_R do
    if not visited [v] then
        if IsBoundary( v) then
            r ← GetLabel( v);
            h ← begin ← FindFirstClockwiseBoundary( v);
            repeat
                Push( β[r], h);
                visited [ Vertex( h) ] ← true;
                h ← FindNextClockwiseBoundary( h);
                h ← Opposite( h);
            until h = begin;
        end
    end
end
```

Figure 6.4: Boundary Traversal Algorithm to identify the vertices at the region boundaries of a segmented mesh. The variable h is a *halfedge* data structure as defined in (Kettner, 1999). The function **Vertex()** returns the vertex to which the halfedge points, and **Opposite()** returns the adjacent halfedge. The **repeat-until** loop walks around the boundary from an initial arbitrary vertex until returning to that same vertex.

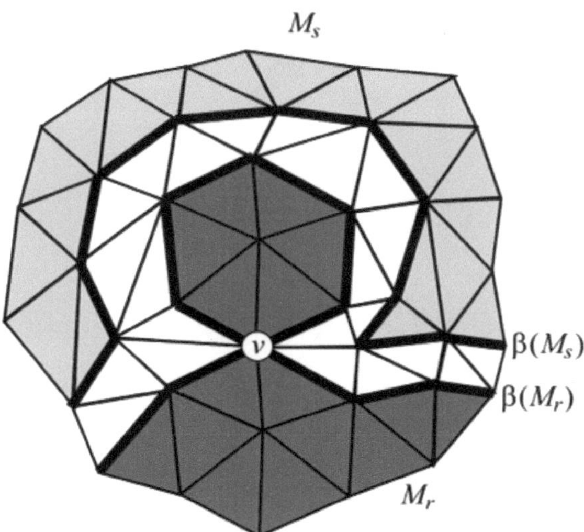

Figure 6.5: A boundary singularity such as v can affect an algorithm that traces the boundary if the algorithm only considers vertices on the boundary and not the topology of the boundary itself. Shaded triangles are the segmentation regions M_r and M_s while white triangles are between regions. Thick lines denote the boundary edges $\beta(M_r)$ and $\beta(M_s)$ of each region.

this function, we begin at the input h and circulate clockwise around v until we reach a halfedge on the boundary that points to v. This time we are walking outside the region across the white triangles as opposed to inside as with the function `FindFirstClockwiseBoundary()`. With this second function, the algorithm moves in a leap-frog fashion around the vertices of a particular boundary and adds those vertices to $\beta(M_r)$.

After identifying the boundaries $\beta(M_R)$, our task of creating graph G is now trivial. We first create R nodes to represent the parts. Then we loop through each $\beta(M_r) \subset \beta(M_R)$, and add edge pairs (r, s) to the graph if $\beta(M_r)$ and $\beta(M_s)$ are connected. Thus, we have an adjacency graph G that compactly describes the segmentation M_R. For visualization, we use a graph drawing toolkit from AT&T Research Laboratories known as *Graphviz*, which lays out drawings of graphs, approaching the quality of manual layouts (Gansner and North, 2000). Our choice of Graphviz results from the fact that automatic layout of graphs is not trivial when they exceed five or ten nodes. The simple examples in the bottom rows of Figs. 6.1 and 6.2 show results from Graphviz.

Other than visualization, we do not directly use this Part Adjacency Graph. However, a graph representation is useful for other computer vision tasks such as object recognition where part decomposition serves as an initial processing step (Trucco and Verri, 1998). Also, a graph representation allows us to apply

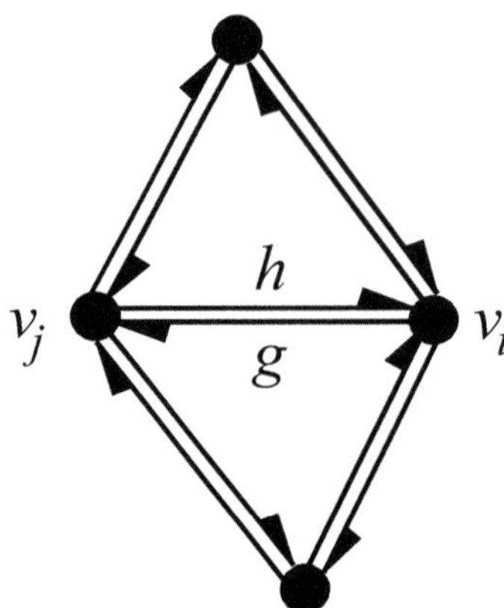

Figure 6.6: An example of the halfedge data structure. An edge (v_i, v_j) in the triangle mesh has two halfedges h and g such that h points to v_i and g to v_j where g is opposite of h. Notice that the halfedges are oriented counterclockwise around a triangle face.

a very rich set of algorithms from graph theory to manipulate our part decomposition. As an example in the next subsection, we explore the side topic of the four-color problem—which is well known in graph theory—for coloring our segmentation regions.

6.2.2 Color Label Selection

With a graph representation such as G, we can easily apply a wide range of graph algorithms to manipulate our segmentation. In particular, a very practical problem that we face when visualizing our segmentation is the choice of colors to label each region. The simplest solution is to choose R unique colors for each of the R parts. This approach works well when our segmentation appears in full color such as on a computer display. However, when it appears in grey scale such as from a laser printer or photocopier, R distinct colors may not necessarily translate into R distinct grey levels, and more importantly a human observer may not necessarily be able to distinguish the grey levels. Our problem now becomes how many grey levels can a human differentiate. As with most psychophysical questions, this one is open for debate and no set number exists. The general consensus is that most people can discriminate at any one point in a monochrome image about one to two dozen different grey levels (Gonzalez and Woods, 2002). For our purposes, we seek sharp contrast among the grey levels so half this number is probably more realistic for coloring our segmentations. Thus, we should probably use about five to ten grey levels. If our number of parts R is greater than five or ten, which is most likely the case, we have a problem since we must repeat some of the grey levels for certain regions. The trick is that we must avoid selecting the same grey level for adjacent parts because those two parts would erroneously appear as a single part. This situation leads to the classical *four-color conjecture*.

In 1852, Francis Guthrie conjectured that only four colors are necessary to color a planar map divided into countries such that two adjacent countries—countries that share a common border—have different colors (Chartrand, 1977). This conjecture is a long-standing problem in graph theory where a map can be represented as an adjacency graph. In 1976, Appel and Haken (Appel and Haken, 1977a; Appel and Haken, 1977b) proffered a proof to this conjecture, but their methodology is a computer assisted approach that has led to skepticism among many mathematicians. Regardless, a rigorous proof does exist for the less constrictive five colors (Heawood, 1890). Heawood explored other topologies beyond the planar map and conjectured the *Heawood number* as the number of necessary and sufficient map colors for a compact connected surface without boundary. For example, the Heawood number for a torus is seven. Heawood's work and the four-color conjecture justify our effort to color our segmentations with a finite set of five to ten grey values. The advantage of representing these segmentations as a graph is that a variety of graph algorithms are available for establishing a four coloring of a graph. The Boost Graph Library that we use with our implementation offers such algorithms, and in particular, we have selected a greedy algorithm that works well with our non-planar topologies to approach the Heawood number (Siek et al., 1999; Siek et al., 2001).

Our brief excursion into this problem is to color the parts of our segmentation for improved visualization. We illustrate the need for intelligent coloring in Fig. 6.7. This example shows the sole of a tennis shoe and is purposefully over segmented to illustrate the coloring problem. In this figure, the top row is two different color maps for the same segmentation. The images immediately below each one are the same coloring except that we have mapped the colors to an equivalent grey scale.* The left column of images shows the result of coloring each region with a simple algorithm that sequentially alternates among a set of colors. The right column shows the more intelligent approach using a four-color greedy algorithm. Notice for Fig 6.7(a) that the same color repeats for the regions across the arch and through the heel of the shoe. These regions are adjacent and thus lead to some confusion. We avoid this problem with the four-coloring algorithm in Fig. 6.7(b). The grey scale images on the bottom row further highlight the problem. The four-color algorithm creates a distinct labeling of the regions and clarifies the visualization of the segmentation, especially for Fig 6.7(d).

---

*The top and bottom rows may appear identical—and in fact are identical—if this document is a non-color photocopy of the original color document.

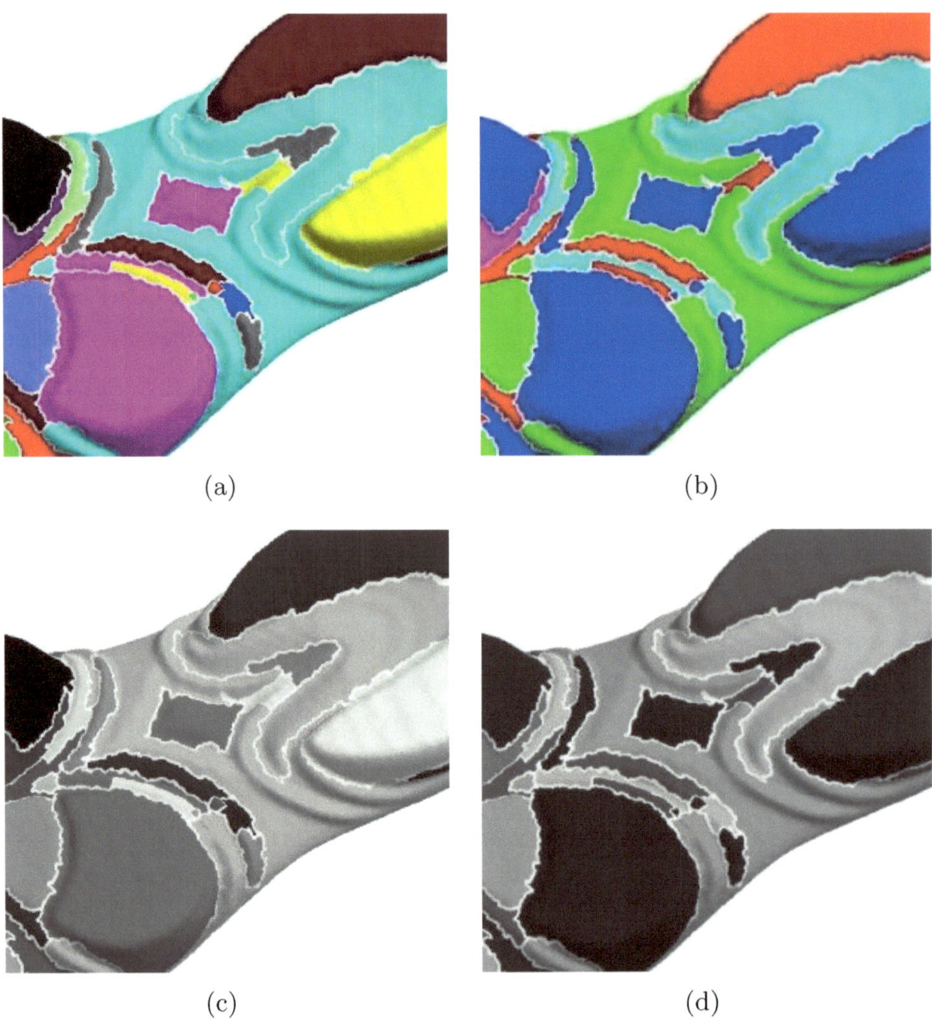

Figure 6.7: This set of segmentations shows possible colorings for the bottom of a tennis shoe. The segmentations are identical and have the same 30 regions. The overall segmentation is intentionally poor to increase the number of adjacent regions to illustrate the four-color problem. (a) This labeling of the regions alternates sequentially among a palette of 16 different colors. (b) This labeling has only four colors for each region such that adjacent regions never share a common color. (c) Grey scale version of (a). (d) Grey scale version of (b). The grey scale versions further motivate the interest in the four-color problem as demonstrated in the improvement from (c) to (d).

At this point, we now have a graph representation of our segmentation, and we have a technique for coloring that graph. Our next objective is to develop the equations for the Part Saliency Metric, which is the subject of the next section.

6.3 Saliency Metric

The major contribution of this chapter is the Part Saliency Metric that we develop in this section. Again, we base the development of this metric on the cognitive theory in (Hoffman and Singh, 1997). Hoffman and Singh follow the minima rule for defining part boundaries and present a theory of part salience. They mainly focus on 2D silhouettes and do not provide rigorous formulations in terms of equations for their definitions. They propose that salience is a function of the relative size of a part to the whole object, the degree to which it protrudes, and the strength of its boundaries. In this section, we present equations for each of these factors, and we specifically address their formulation in terms of 3D triangle meshes.

We begin by proposing that the overall salience $\mathcal{S}_r$ of a part r is the weighted sum of the three factors discussed by Hoffman and Singh. Given the relative size of a part $\mathcal{S}_\sigma$, the degree of protrusion $\mathcal{S}_\pi$, and the strength of the boundary $\mathcal{S}_\beta$, we can write the following equation:

$$\mathcal{S}_r = \frac{1}{3}\left(w_\sigma \mathcal{S}_\sigma + w_\pi \mathcal{S}_\pi + w_\beta \mathcal{S}_\beta\right) \tag{6.2}$$

where w_σ, w_π, and w_β are the weights of each factor, respectively, such that

$$w_\sigma + w_\pi + w_\beta = 1 \, . \tag{6.3}$$

This formulation insures the salience $\mathcal{S}_r$ is in the range $0 \leq \mathcal{S}_r \leq 1$. The constraint in (6.3) means that the user only needs to select two of the three weights since the third one follows. Two of the weights, however, are indeed free parameters and are necessary since it is unclear as to the relative significance of each factor to the overall perception of part salience (Hoffman and Singh, 1997). More experimental research from cognitive perception is necessary to understand the inter-relationship among these factors, and so we simply leave these weights as user parameters. We do not investigate perceptual significance of these weighting terms in this dissertation. We leave this question for future research and specifically to future cognitive research. The block diagram in Fig. 6.8 illustrates our methodology. In each of the next three subsections, we discuss how to compute each factor: $\mathcal{S}_\sigma$, $\mathcal{S}_\pi$, and $\mathcal{S}_\beta$.

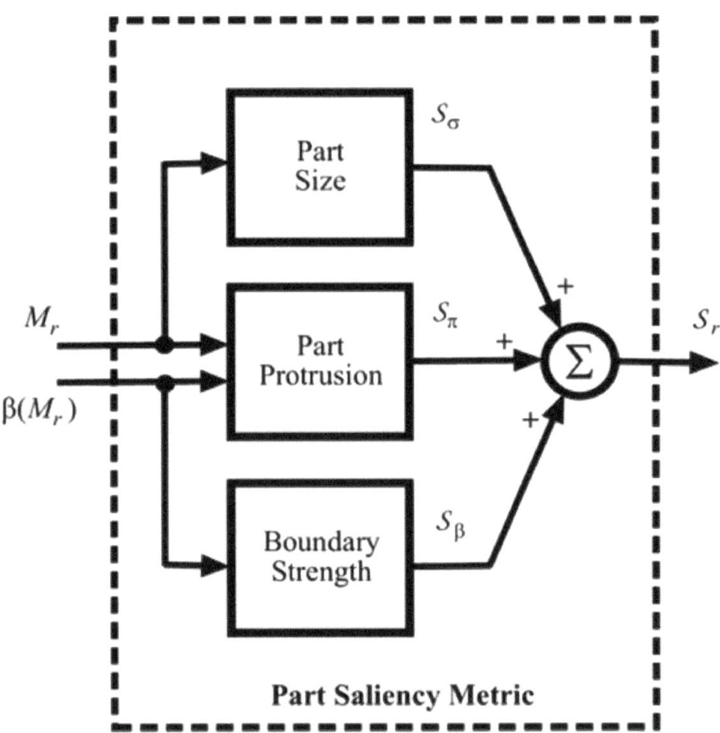

Figure 6.8: Block diagram of the Part Saliency Metric. The input is a part M_r and the boundary $\beta(M_r)$ for the part. The output is the salience of the part S_r.

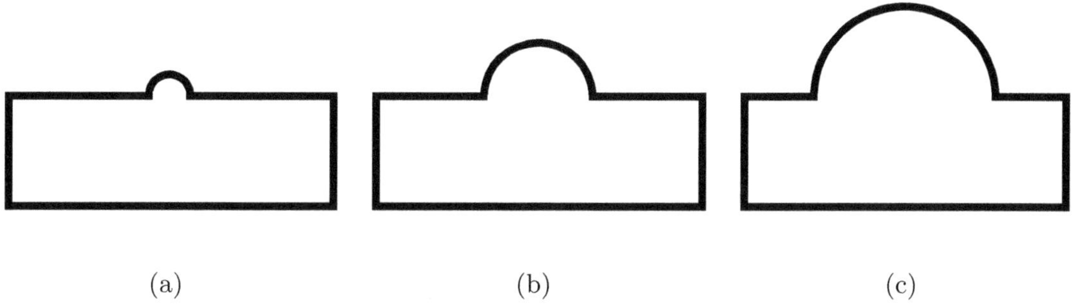

(a)   (b)   (c)

Figure 6.9: This sequence shows how increasing part sizes corresponds to increasing salience of the part. The 2D silhouettes show an object with two parts: a circle and a square. The circle increases in size from left to right.

6.3.1 Part Size

The first factor we explore is the relative size measure $\mathcal{S}_\sigma$ of a part. This measure reflects the volume of the part relative to the volume of whole object. See Fig. 6.9. For our application, this measure is the volume bounded by the part M_r relative to the volume bounded by the original mesh M. Unfortunately, computing the volume of either M_r or M is not straightforward since these meshes are possibly open surfaces, and by definition, an open surface does not bound a volume. The part M_r may be an open surface since we may not necessarily know how to complete the part itself, particularly across the boundary contours $\beta(M_r)$. A part-cut rule such as (Singh et al., 1999) or (Rosin, 1999) is necessary but part-cut theories are not well understood in terms of human perception and thus difficult to formulate in terms of computer algorithms. To demonstrate the ambiguity of part cuts, consider the simple illustration in Fig. 6.10. As for the original mesh M, it should ideally be a closed surface, but it may have holes from scanning occlusions and other imperfections that preclude it from being a closed mesh, as well. For these reasons, we suggest that the bounding boxes of these meshes are an appropriate approximation of their volumes, and we thus suggest the following:

$$\mathcal{S}_\sigma = \frac{B_r}{B_M} \tag{6.4}$$

where $0 \leq \mathcal{S}_\sigma \leq 1$, B_r is the volume of the bounding box of the part M_r, and B_M is the volume of the bounding box for the entire mesh.

The question we now face is how to compute the bounding box of an open mesh M_r. A variety of solutions are possible (O'Rourke, 1994). We propose a method based on the scatter matrix of the vertices of M_r and an eigen analysis of that matrix. The scatter matrix $\mathbf{S}_\sigma$ is simply the covariance matrix of the

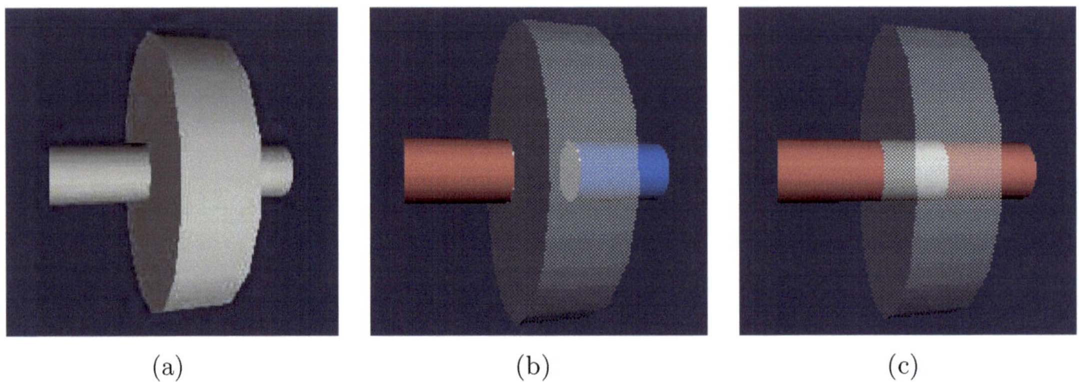

Figure 6.10: These images show the ambiguity associated with part cuts. (a) A simple object that consists of a wheel with an axel. (b) If we chose vertical part cuts, then the decomposition consists of three parts: a wheel and a left and right axel peg. (c) If we choose a single horizontal part cut, then we have only two parts: a wheel and an axel.

vectors s_i where we define s_i as

$$s_i = v_i - \bar{v}_i \tag{6.5}$$

and the average vector $\bar{v}_i$ as

$$\bar{v}_i = \frac{1}{n_r} \sum_{v_i \in M_r} v_i \, . \tag{6.6}$$

The 3D vectors $v_i \in M_r$ are the vertices of the mesh for part r and n_r is the number of such vertices. We formulate the 3×3 scatter matrix as the outer product of the s_i vectors as

$$\mathbf{S}_\sigma = \sum_{v_i \in M_r} s_i s_i^t \tag{6.7}$$

where the summation is over the set of s_i vectors and t denotes the transpose. To compute the bounding box volume of M_r, we perform an eigenvalue decomposition of $\mathbf{S}_\sigma$, which yields the three eigenvalues λ_1, λ_2, and λ_3. We then estimate the volume as the product of these three values

$$B_r = 8\sqrt{\lambda_1 \lambda_2 \lambda_3} \, . \tag{6.8}$$

From these values, we estimate the volume B_M of the whole mesh as the sum of each part,

$$B_M = \sum_r B_r \, . \tag{6.9}$$

The value is more accurate than the bounding volume of the entire mesh. The next step is compute the degree of protrusion of the part.

6.3.2 Part Protrusion

The second measure that we must compute is the degree of protrusion. This measure is the degree to which a part "sticks out" from its object (Hoffman and Singh, 1997). Parts that stick out more seem to be more salient. See Fig. 6.11. A couple of definitions for protrusion are possible (Siddiqi and Kimia, 1995; Hoffman and Singh, 1997). Hoffman and Singh argue in a qualitative manner that the most appropriate is the ratio of the surface area of the part to the area of its cropped base. The cropped base is the area of the surface bounded by the contour of the part boundary $\beta(M_r)$ that cuts through the object to distinguish the part from the object. Recall Fig. 6.10 and the ambiguity of part cuts. For now, we assume we know what part cut and cropped base we have. We write the following equation to formalize the relative protrusion $\mathcal{S}_\pi$ of a part r.

$$\mathcal{S}_\pi = 1 - \frac{A_\beta}{A_r} \tag{6.10}$$

where $0 < \mathcal{S}_\pi \leq 1$, A_r is the surface area of the part, and A_β is the surface area of the cropped base. The subtraction from one insures that this function increases as the part protrudes more from the object. The computation of the part area A_r directly follows from the area of the triangles that compose the surface of the part.

$$A_r = \sum_{f_i \in M_r} A_i \tag{6.11}$$

where A_i is the area of the triangle f_i for part r. Unfortunately, we can not compute the area of the cropped base quite as easily.

The difficulty with the cropped base is that we actually do not know the surface that the boundary $\beta(M_r)$ bounds. We do not have a unique specification of the base of a part and how it joins with the object because we lack an adequate part-cut theory. We only have the part boundary. We could identify a minimal surface where the boundary serves as a constraint along the lines of "Plateau's Problem" (Stuwe, 1989). Unfortunately, a calculus of variations solution makes such an approach unattractive from a computational standpoint. So, our solution is to again turn to a scatter matrix. This time our matrix is only for the vertices v_j that reside along the boundary $\beta(M_r)$ of part r. We compute the 3×3 matrix $\mathbf{S}_\beta$

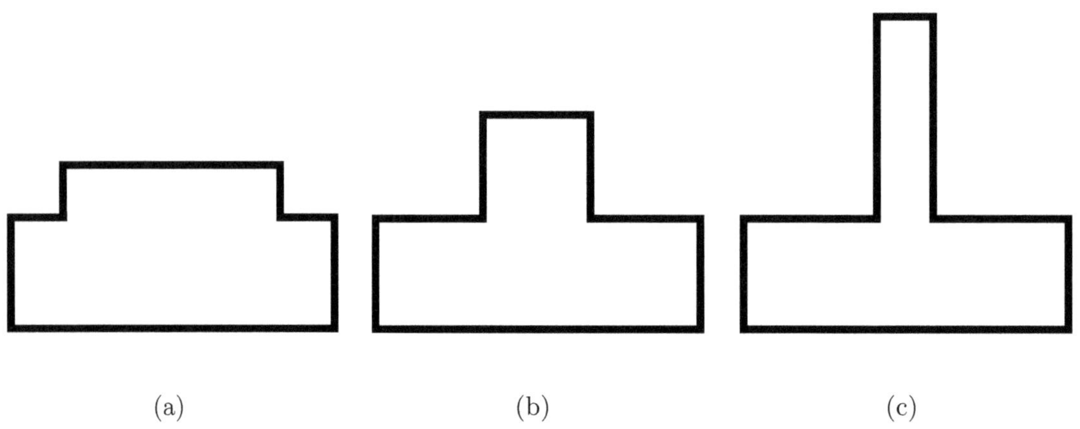

(a)   (b)   (c)

Figure 6.11: This sequence shows how increasing part protrusion corresponds to increasing salience of the part. The 2D silhouettes show an object with two rectangular parts. The top rectangle for each subfigure has the same area, but its protrusion increases from left to right.

as the covariance of the scatter vectors s_j as follows

$$\mathbf{S}_\beta = \sum_{v_j \in \beta(M_r)} s_j s_j^t \tag{6.12}$$

where t denotes transpose. The scatter vectors are defined as

$$s_j = v_j - \bar{v}_j \tag{6.13}$$

where the vector average is

$$\bar{v}_j = \frac{1}{n_\beta} \sum_{v_j \in \beta(M_r)} v_j \; . \tag{6.14}$$

The number n_β is the number of vertices v_j in the boundary $\beta(M_r)$. We decompose $\mathbf{S}_\beta$ into the eigenvalues λ_1, λ_2, and λ_3. If we assume that the boundary contour $\beta(M_r)$ is close to—but not necessarily—planar, we can now estimate the area A_β of the cropped base as the product of the two largest eigenvalues.

$$A_\beta = 4\sqrt{\lambda_1 \lambda_2} \tag{6.15}$$

where $\lambda_1 \geq \lambda_2 \geq \lambda_3 \geq 0$ since $\mathbf{S}_\beta$ is a semi-definite symmetric matrix. This equation means that we indirectly find A_β from the bounding box of the part boundary. We find the two largest dimensions λ_1 and λ_2 of that box and compute the area from those dimensions. The assumption is that the third dimension λ_3 is nearly zero since we assume the boundary is almost planar. Most part boundaries should not violate this simple assumption. As a caveat, if a part contains more than one boundary, we use the boundary with the largest estimated area in the numerator of Eq. (6.10). We add the areas of the other boundaries for the part back to A_r. This process improves the estimate of A_r almost as if we are filling in the wholes formed by parts extending from the current part. Our next task is to compute the boundary strength.

6.3.3 Boundary Strength

The final measure that we need to compute for the part salience is the boundary strength. See Fig. 6.12. Recall that the minima rule defines part boundaries along lines of negative minima curvature. Thus, the strength of the boundary is the degree of curvature along the boundary. One measure of this degree is an average of the minimum principal curvatures for each vertex on the boundary. So, for a part r, we know from the previous section the boundary $\beta(M_r)$ for that part. We can compute this average as follows:

$$\mathcal{S}_\beta = \frac{1}{n_\beta} \sum_{v_j \in \beta(M_r)} \left| \frac{\kappa_j^2}{\kappa_{min}} \right| \tag{6.16}$$

where $0 \leq \mathcal{S}_\beta \leq 1$, n_β is the number of vertices in the boundary set, κ_j^2 is the minimum principal curvature for vertex v_j. On the rare occasion when $\kappa_j^2 > 0$ for a particular v_j, we just omit this value from the summation, but we still include it in the n_β count. We normalize this summation with the most negative curvature κ_{min} from the whole mesh.

$$\kappa_{min} = \min \left\{ \kappa_i^2 \mid v_i \in \beta(M_R) \right\} \tag{6.17}$$

where $\beta(M_R)$ is the set of boundaries for every part from the segmented mesh M_R.

This section completes our salience definitions. We now plug each measure back into Eq. (6.2) to compute the salience of a part. The salience reflects the quality of each part, as a part, and allows us to merge bad parts with other more salient ones. We address this issue in the next section.

6.4 Filter and Merge Algorithm

The previous section allows us to assign a salience value $\mathcal{S}_r$ to each part r of the segmented mesh M_R. With this value, we can rank the importance of each part to evaluate the quality of the segmentation. Assuming that our Part Saliency

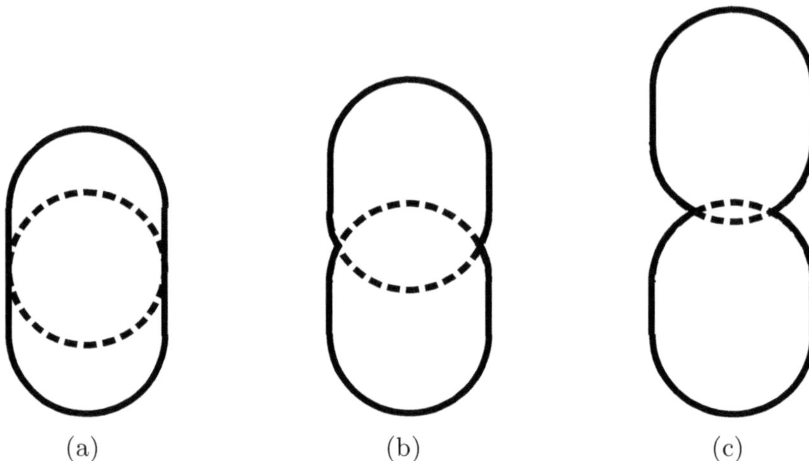

Figure 6.12: This sequence shows how increasing the boundary strength of a part corresponds to increasing salience of the part. The 2D silhouettes show an object with two circular parts. The boundary strength is the degree of negative curvature at the part boundary. For example, the boundary strength is zero—non-existent—for the leftmost figure, which without the dotted lines we perceive as a single part. The boundary strength is strongest for the rightmost figure.

Metric reflects human visual perception, a part with a larger salience value is a more important part, and thus better quality, than one with a smaller value. If we sort the parts into an ascending list with higher values of salience at the top of the list, then "poor" quality parts trickle to the bottom of the list. The last part in the list is the least important part within M_R. With this ordered list, we filter a part at the bottom of the list and merge it with another more salient parts. This process completes the final step in the block diagram of Fig. 6.3 and serves to improve any oversegmentation of M_R that may have occurred.

If we assume that the salience of the last part in our list is such that it is a "bad" part, then we can merge it with one of its neighboring parts. The Part Adjacency Graph, G, tells us which parts are the neighbors of this bad part, and we can use this graph to determine where to merge. If the bad part has only one neighbor, the choice is simple. We merge it with that neighbor and then recompute the salience for the subsequent combined part. We reinsert this new part and its salience into the ranking appropriately. If, however, the bad part has more than one neighbor, we must choose which one is more suitable for a merge. Hoffman and Singh do not discuss this issue and thus do not suggest theories of human perception. Without an adequate theory, we suggest the relative salience of each neighboring part governs this choice. If a part has two or more adjacent neighbors, we select the least salient neighbor as the candidate for merging. As before, we merge these two parts into a single one, compute their resulting combined salience and insert this new part into the ordered list.

With this filter-and-merge process, we are able to use the salience of a part to improve the decomposition of an object (or scene) from our watershed segmentation of the previous chapter. The key to this process is the definition of a good and bad part. At the current stage of this research, we do not suggest an absolute definition for these terms, and indeed such a definition may not exist, even for human perception. We do, however, propose a computational framework that applies the theories of Hoffman and Singh to triangle mesh representations that are common in computer vision. Our framework assigns to each part r a salience value $\mathcal{S}_r$ where a threshold of this value may serve as a definition for good and bad parts. The difficulty is that if we merge part r with part s, they form a new part r' with its own salience $\mathcal{S}_{r'}$. Thus, a straight threshold of the aggregate of parts from M_R to establish a set of good and bad parts is not possible. Rather an iterative, trimming threshold is necessary as our process above outlines. We order the parts in a list and trim the least salient one from the bottom of the list. Then, we reevaluate the updated list that includes the merged part to determine if another bad part is at the bottom.

6.5   Remarks

Our final remark for this chapter concerns the user parameters for the equations discussed in the previous sections. To compute $\mathcal{S}_r$ for each part, the user must select two of three weights to define the relative significance of each component of the Part Saliency Metric. Table 6.1 outlines each of these parameters. The intent of these parameters is to provide a means for a user to mimic human vision preferences for each component of salience. When a user needs to place more emphasis on one component of salience, she simply increases the weight for that component. For example, with industrial parts such as bolts, screws, and other fabricated components, the boundary strength may not be as important as part size and protrusion since these objects typically have well-defined boundaries that are often right angles. Since the strength of such part boundaries are almost equivalent, they do not play a major role in the salience of parts.

The Part Saliency Metric in this chapter along with Normal Vector Voting and Fast Marching Watersheds from the previous chapters are the major steps of our total part decomposition algorithm. In the next chapter, we first explore results from each of the algorithms individually. Then we combine them into a complete system and explore experimental results with data sets of real-world objects and scenes from a variety of range scanning devices.

Table 6.1: Parameters for Part Saliency Metric. Eq. (6.3) constrains these three parameters such that the user selects two and the third follows.

Parameter	Range	Equation	Typical Value	Comments
w_σ	$0 \le w_\sigma \le 1$	(6.2)	$w_\sigma = 0.5$	Relative importance of part size to the salience of the part.
w_π	$0 \le w_\pi \le 1$	(6.2)	$w_\pi = 0.5$	Relative importance of part protrusion to the salience of the part.
w_β	$0 \le w_\beta \le 1$	(6.2)	$w_\beta = 0.1$	Relative importance of boundary strength to the salience of the part.
Recall Eq. (6.3) $w_\sigma + w_\pi + w_\beta = 1$.				

Chapter 7

Experimental Results

This chapter presents the experimental results for the algorithms proposed in this dissertation. We begin with results from the Minima Rule Algorithm in Sec. 7.1. This section demonstrates the overall capabilities of the algorithm relative to human visual perception through the theory of the minima rule. This first section is a high-level presentation of the results and does not discuss in detail the strengths and weaknesses of the algorithm. The next three sections, however, do provide a thorough analysis of the individual components of the algorithm. First, in Sec. 7.2, we show the robustness of the Normal Vector Voting algorithm for the estimation of surface curvature. Then, in Sec. 7.3, we investigate the capabilities of the Fast Marching Watersheds algorithm to identify the minima rule boundaries and decompose a mesh into visual parts. Finally, in Sec. 7.4, we evaluate the Part Saliency Metric as a measure of the quality of the results. These four sections serve as the successful evidence of our proposed algorithms.

7.1 Minima Rule Algorithm

The Minima Rule Algorithm decomposes triangle meshes into visual parts. In previous chapters, we have described the motivation for this algorithm and we have outlined the theory that supports the development of the algorithm. In this section, we present the results from our implementation.

We have coded the complete Minima Rule Algorithm in Visual C++® on a Microsoft Windows platform using OpenGL for 3D visualization. A screen shot of the interface for the program appears in Fig. 7.1. This main window of the program is a 3D mesh viewer with a trackball controller. The menu and buttons above the visualization window are the user interface for each aspect of the Minima Rule Algorithm such as the curvature estimation, the threshold selection, the morphology operations, and other functions previously outlined. We have

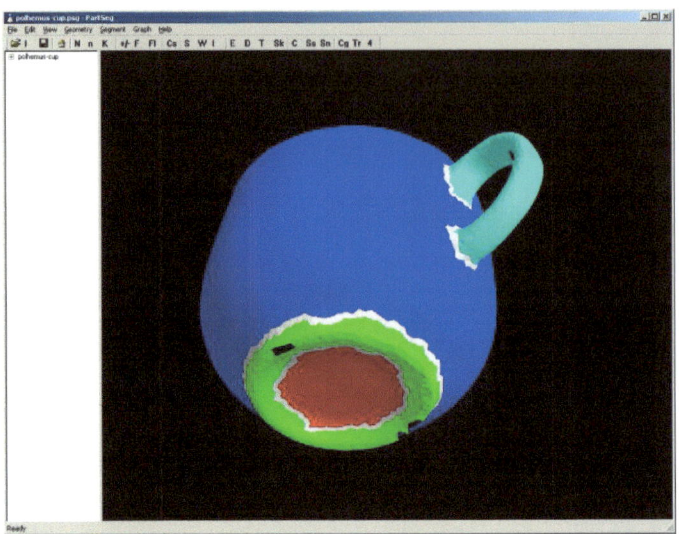

Figure 7.1: Screen shot of user interface for Minima Rule Algorithm program.

implemented these functions in standard C++ (Weiss, 1999; Stroustrup, 1991) using a variety of libraries as tabulated in Table 7.1.

Using this program, we can decompose a mesh that is a manifold surface and has arbitrary topology into minima rule parts. To demonstrate this capability, we show the decomposition results for a broad spectrum of meshes in Figs. 7.2 through 7.23. Each figure consists of a rendering of the particular object or scene, the color-coded segmentation of that object or scene, and an adjacency graph of the segmentation parts. In some cases, we have also included a photograph of the actual object or scene. Table 7.2 briefly describes each mesh in the figures. We have two fundamental sources for each mesh. We have either generated the meshes in-house using our range scanners and reconstruction algorithms as outlined in (Sun et al., 2002b; Page et al., 2003d) or used meshes from external sources such as other laboratories or commercial companies. The previous table designates the source of each mesh, and Tables 7.3 and 7.4 give additional details about each source. Table 7.3 summarizes the types of scanners and their accuracy that we used to generate in-house reconstructions. On the other hand, Table 7.4 documents the external sources of meshes.

We present these figures and results without much discussion. We delay detail analysis of the Minima Rule Algorithm to the next sections where we individually investigate each component of the algorithm. This section merely serves as a broad overview of the results. For this overview, one should view each rendering of the meshes throughout the figures and mentally decompose them into parts. Then, one should compare this mental image with the color-coded decompositions in the figures. This comparison should yield strong agreement between one's mental selection of parts and the Minima Rule Algorithm's selection of parts. We do not suggest that total agreement is possible

Table 7.1: List of coding libraries used in software development.

Library Name	Description	Technical Reference
Microsoft Foundation Classes	Windows interface library	(White et al.,)
OpenGL	3D graphics visualization library	(Neider et al., 1993)
Standard Template Library	flexible and extensible set of software building blocks	(Hughes and Hughes, 1999)
Computational Geometry Algorithms Library	geometry and mesh manipulation algorithms	(Veltkamp, 1999)
Numerical Recipes	numerical computation algorithms	(Press et al., 1992)
Boost Graph Library	graph creation and manipulation algorithms	(Siek et al., 1999)
Graphviz	graph visualization algorithms	(Gansner and North, 2000)

Table 7.2: List of triangle meshes for Minima Rule Algorithm results. The last column shows the number of parts after the mesh decomposition.

Figure Number	Brief Description	Date Source	Vertex Count	Triangle Count	Part Count
7.2	hand crank	RANGER	46, 870	93, 752	7
7.3	water neck	RANGER	58, 784	117, 564	6
7.4	distributor cap	RANGER	65, 397	129, 849	20
7.5	disc brake	RANGER	37, 332	73, 553	2
7.6	bin objects	RANGER	39, 752	71, 976	42
7.7	cone scene	COLEMAN	61, 027	117, 778	6
7.8	office scene (I)	RIEGL	65, 365	129, 342	39
7.9	office scene (II)	PERCEPTRON	15, 666	30, 981	28
7.10	bore pin	TACOM	37, 450	74, 896	4
7.11	toilet	POLHEMUS	22, 926	45, 864	5
7.12	watering can	POLHEMUS	8, 086	15, 843	6
7.13	chair	POLHEMUS	26, 766	53, 462	11
7.14	left hand	POLHEMUS	13, 340	26, 372	8
7.15	oil pump	HOPPE	19, 555	39, 102	15
7.16	teapot	HOPPE	3, 034	6, 010	5
7.17	human femur	ITALY	76, 794	153, 322	2
7.18	machined object	SLIM-3D	28, 667	57, 107	10
7.19	human molar tooth	SLIM-3D	6, 586	13, 168	4
7.20	human canine tooth	SLIM-3D	3, 376	6, 748	2
7.21	fan blades	3D DIGITAL	121, 271	239, 227	16
7.22	shoe sole	3D DIGITAL	58, 108	115, 750	18
7.23	David's head	STANFORD	7, 790	15, 203	20

Table 7.3: List of range scanners used for in-house mesh reconstructions. The National Automotive Center is with the U. S. Army Tank and Automotive Command (TACOM). A technical reference for this TACOM data set is not available at this time. The abbreviations *Acc.* and *Max.* each mean approximate accuracy and maximum range, respectively.

Scanner Label	Manufacturer	Type	Acc. (cm)	Max. (m)	Technical Reference
RANGER	Integrated Vision Products	Active Stereo	0.1	0.2	(Integrated Vision Products, 2000)
COLEMAN	Coleman Research Corporation	Time of Flight	0.1	20	(Sebastian et al., 1995)
RIEGL	Riegl Laser Measurement Systems	Time of Flight	5	700	(Riegl Laser Measurement Systems, 2000)
PERCEP-TRON	Perceptron Incorporated	Time of Flight	5	10	(Perceptron Incorporated, 1993)
TACOM	National Automotive Center	Laser Probe	0.01	20	N/A

Table 7.4: List of external sources for triangles meshes used in results.

Source Label	Company	Technical Reference
POLHEMUS	Applied Research Associates	(McCallum et al., 1998)
HOPPE	Microsoft Research	(Hoppe et al., 1992)
ITALY	Istituti Ortopedici Rizzoli	(Viceconti et al., 1996)
SLIM-3D	Friedrich-Alexander-Universität Erlangen	(Karchaher et al., 1997)
3D DIGITAL	3D Digital Corporation	(3D Digital Corporation, 2000)
STANFORD	Stanford University	(Levoy et al., 2000)

since a human observer uses other cognitive strategies to select parts such as context and experience. However, these results should not totally contradict one's perception either. This qualitative exercise should provide convincing evidence of the capabilities of the algorithm.

These figures serve as a demonstration of the Minima Rule Algorithm. We now turn to a more in-depth analysis of the capabilities of the algorithm. In the next three sections, we investigate the Normal Vector Voting algorithm, the Fast Marching Watersheds algorithm, and the Part Saliency Metric algorithm.

7.2   Normal Vector Voting

The heart of the Minima Rule Algorithm is the Normal Vector Voting algorithm that estimates the orientation and curvature of the triangle mesh where we assume the mesh approximates some smooth surface. To demonstrate the capabilities of this algorithm, we have experimented with a variety of data sets from CAD models to range reconstructions to medical isosurfaces. In this subsection, we present both the qualitative results of these experiments and a quantitative comparison from a set of more controlled experiments. In the discussions that follow, we must recall the user parameters for the algorithm as outlined in Table 4.2. The three parameters are: k the neighborhood size, ε the crease detection constant, and η the noise suppression constant. We mainly specify k, but when pertinent we also call out the system constants ε and η. These latter two constants, in actual practice, are fixed and should not change. The intent is that we control the algorithm with k and not these constants.

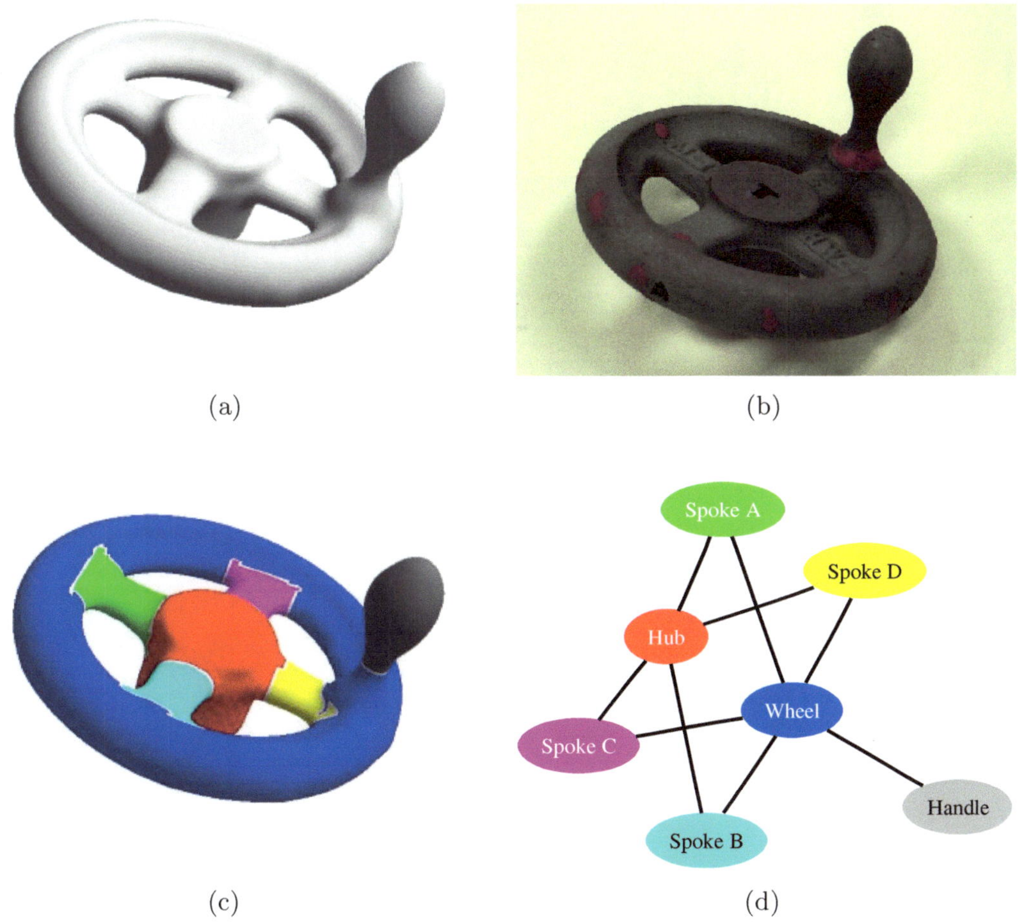

(a)

(b)

(c)

(d)

Figure 7.2: Minima Rule Algorithm results for hand crank object. This mesh is a reconstruction from multiple range scans from the IVP Ranger System. The mesh consists of $46,870$ vertices and $93,752$ triangles while the decomposition consists of 7 parts. (a) Rendering of original mesh. (b) Photograph of original object. (c) Decomposition of mesh into parts. (d) Adjacency graph with user-specified labels.

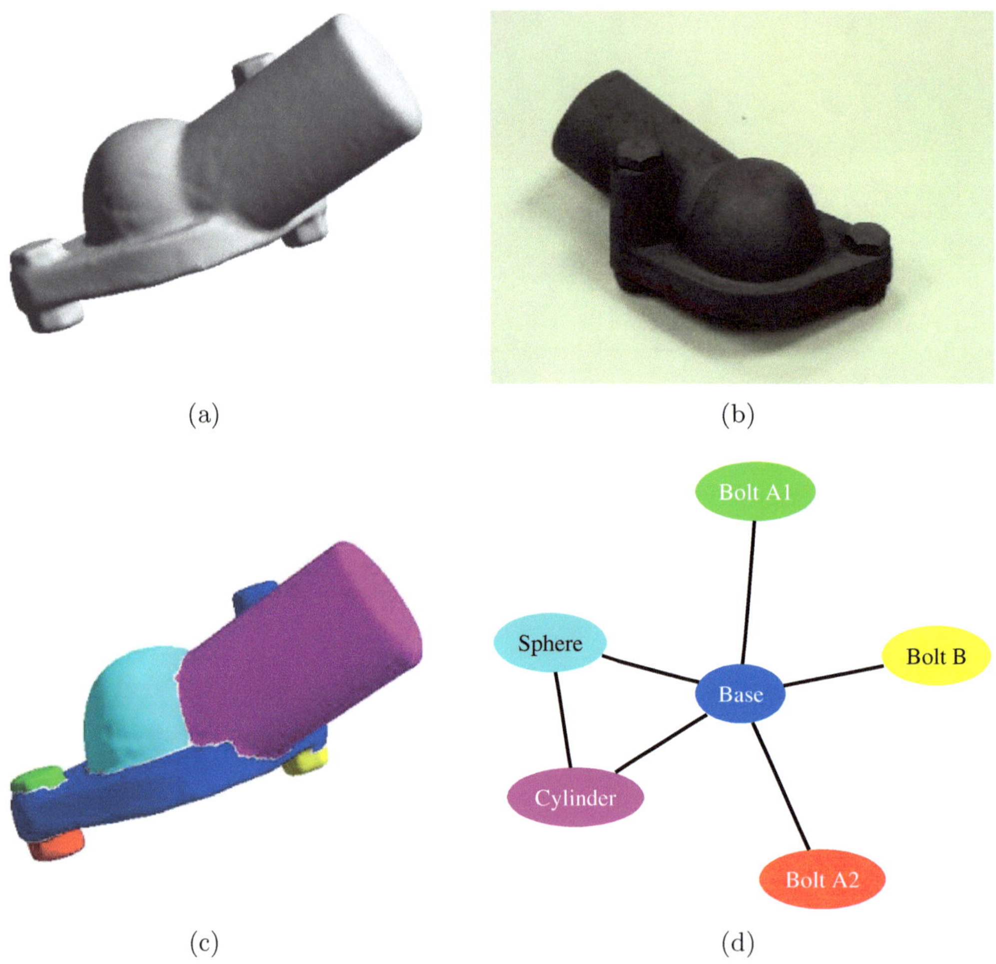

Figure 7.3: Minima Rule Algorithm results for water neck object. This mesh is a reconstruction from multiple range scans from the IVP Ranger System. The mesh consists of $58,784$ vertices and $117,564$ triangles while the decomposition consists of 6 parts. (a) Rendering of original mesh. (b) Photograph of original object. (c) Decomposition of mesh into parts. (d) Adjacency graph with user-specified labels.

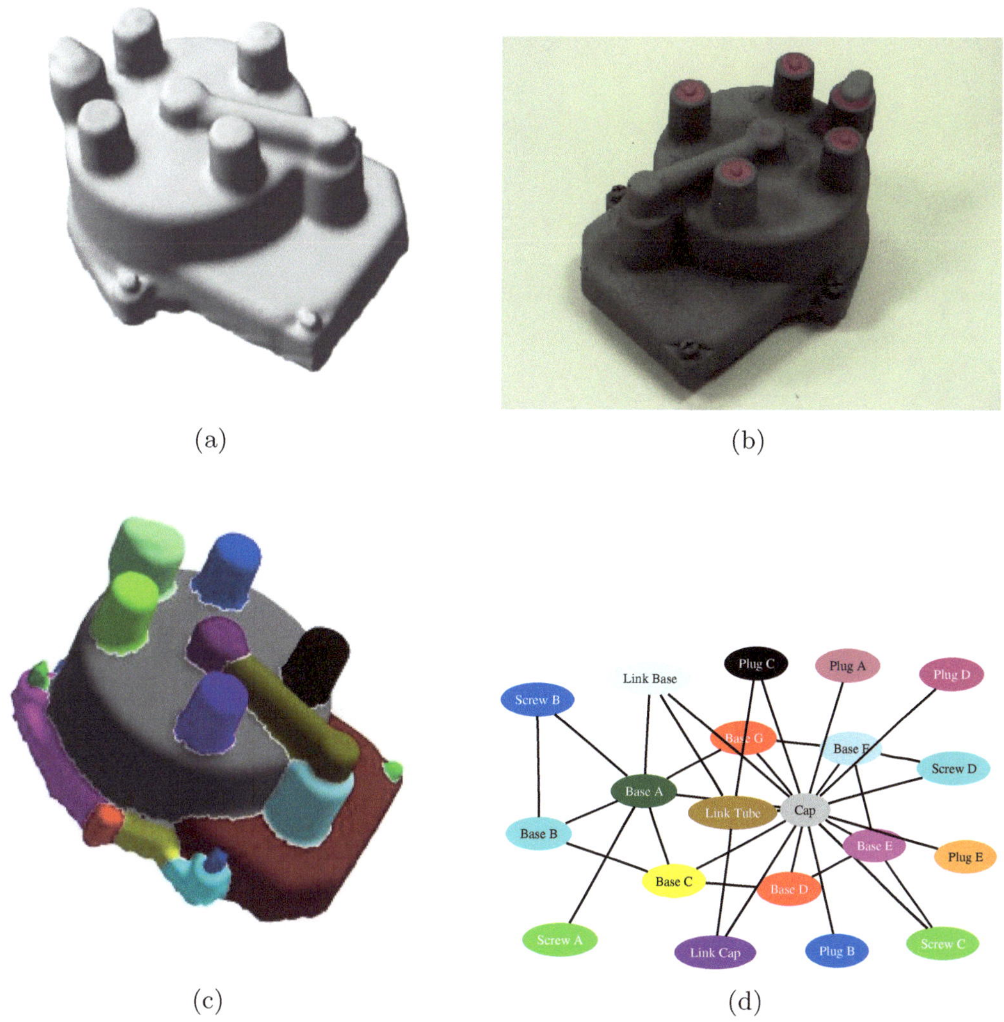

Figure 7.4: Minima Rule Algorithm results for distributor cap object. This mesh is a reconstruction from multiple range scans from the IVP Ranger System. The mesh consists of $65,397$ vertices and $129,849$ triangles while the decomposition consists of 20 parts. (a) Rendering of original mesh. (b) Photograph of original object. (c) Decomposition of mesh into parts. (d) Adjacency graph with user-specified labels.

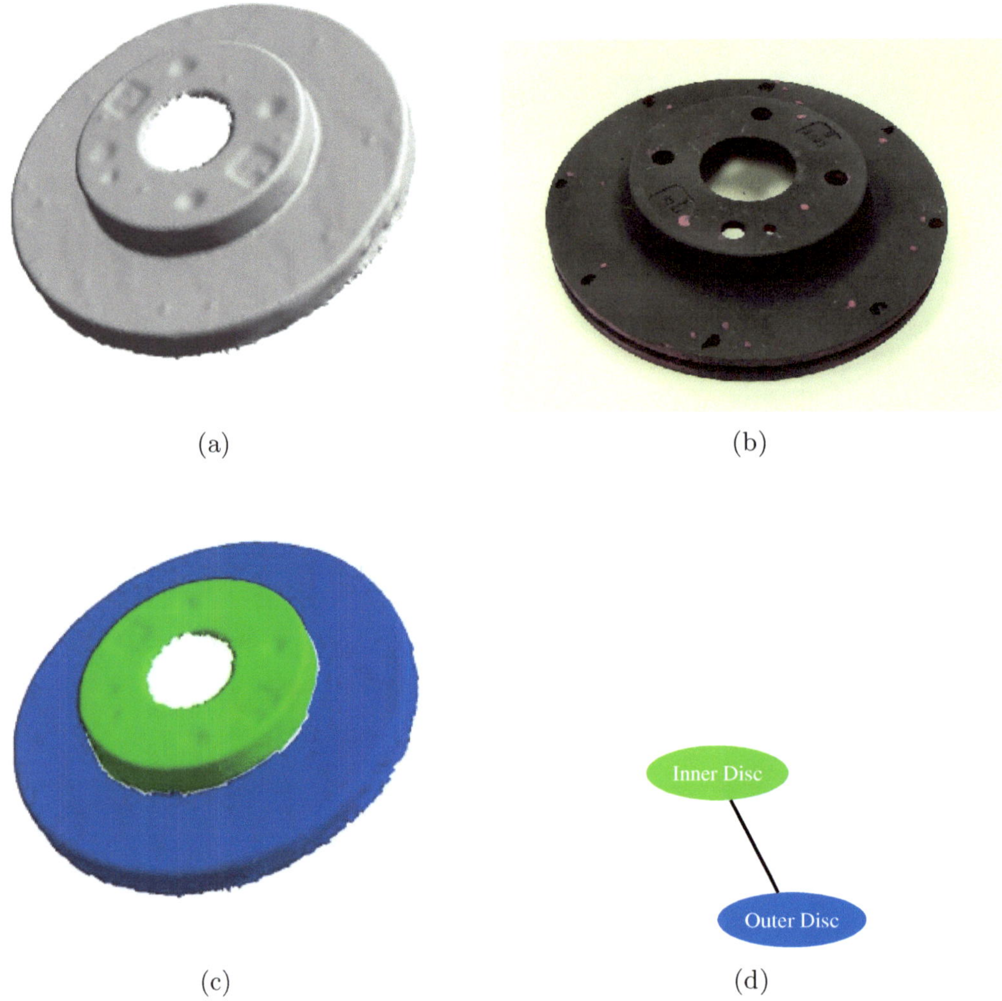

(a)

(b)

(c)

(d)

Figure 7.5: Minima Rule Algorithm results for disc brake object. This mesh is a reconstruction from multiple range scans from the IVP Ranger System. The mesh consists of $37,332$ vertices and $73,553$ triangles while the decomposition consists of 2 parts. (a) Rendering of original mesh. (b) Photograph of original object. (c) Decomposition of mesh into parts. (d) Adjacency graph with user-specified labels.

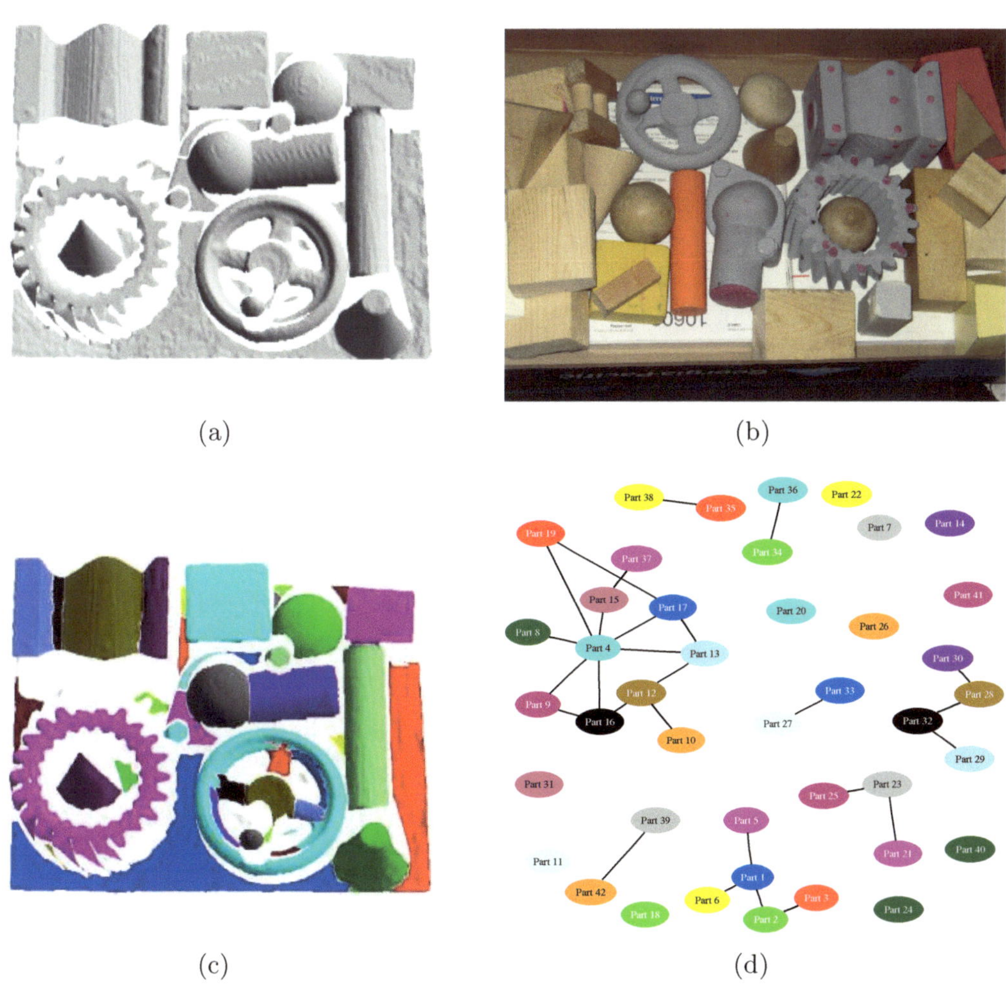

(a)   (b)

(c)   (d)

Figure 7.6: Minima Rule Algorithm results for miscellaneous objects simulating a bin picking application. This mesh is a single view from the IVP Ranger System. The mesh consists of $39,752$ vertices and $71,976$ triangles while the decomposition consists of 42 parts. Since this mesh is a single view, the scan leaves many objects as isolated parts and thus unconnected to other surfaces. (a) Rendering of original mesh. (b) Photograph of approximate scene. (c) Decomposition of mesh into parts. (d) Adjacency graph with arbitrary labels.

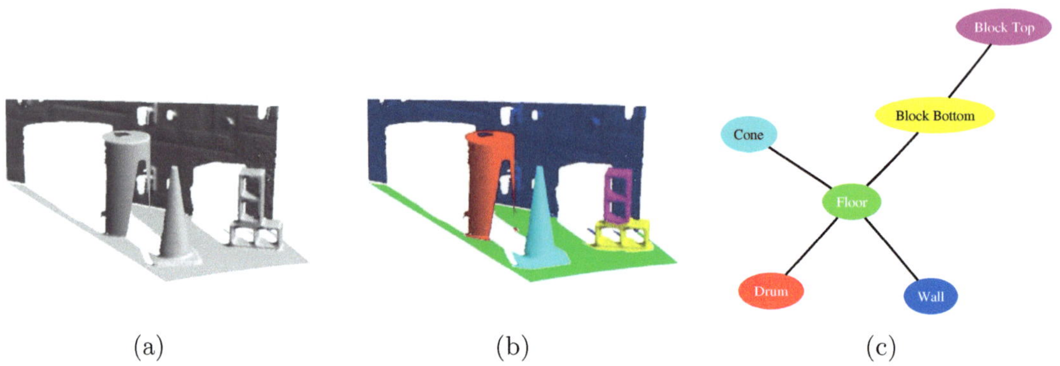

(a)   (b)   (c)

Figure 7.7: Minima Rule Algorithm results for scene with cone and barrel. This mesh is a single view from the Coherent Laser Radar System developed by Coleman Research Corporation. The mesh consists of $61,027$ vertices and $117,778$ triangles while the decomposition consists of 6 parts. (a) Rendering of original mesh. (b) Decomposition of mesh into parts. (c) Adjacency graph with user-specified labels.

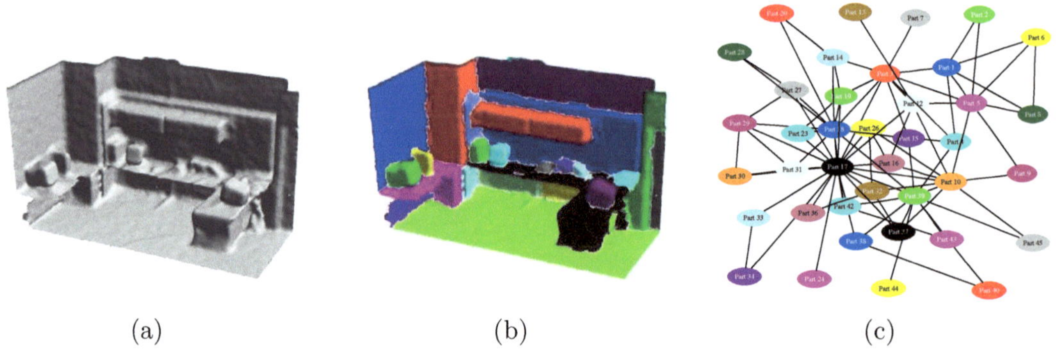

(a)   (b)   (c)

Figure 7.8: Minima Rule Algorithm results for an office scene. This mesh is a reconstruction from multiple range scans from the Laser Mirror System Z210 developed by Riegl. The mesh consists of $65,365$ vertices and $129,342$ triangles while the decomposition consists of 39 parts. (a) Rendering of original mesh. (b) Decomposition of mesh into parts. (c) Adjacency graph with arbitrary labels.

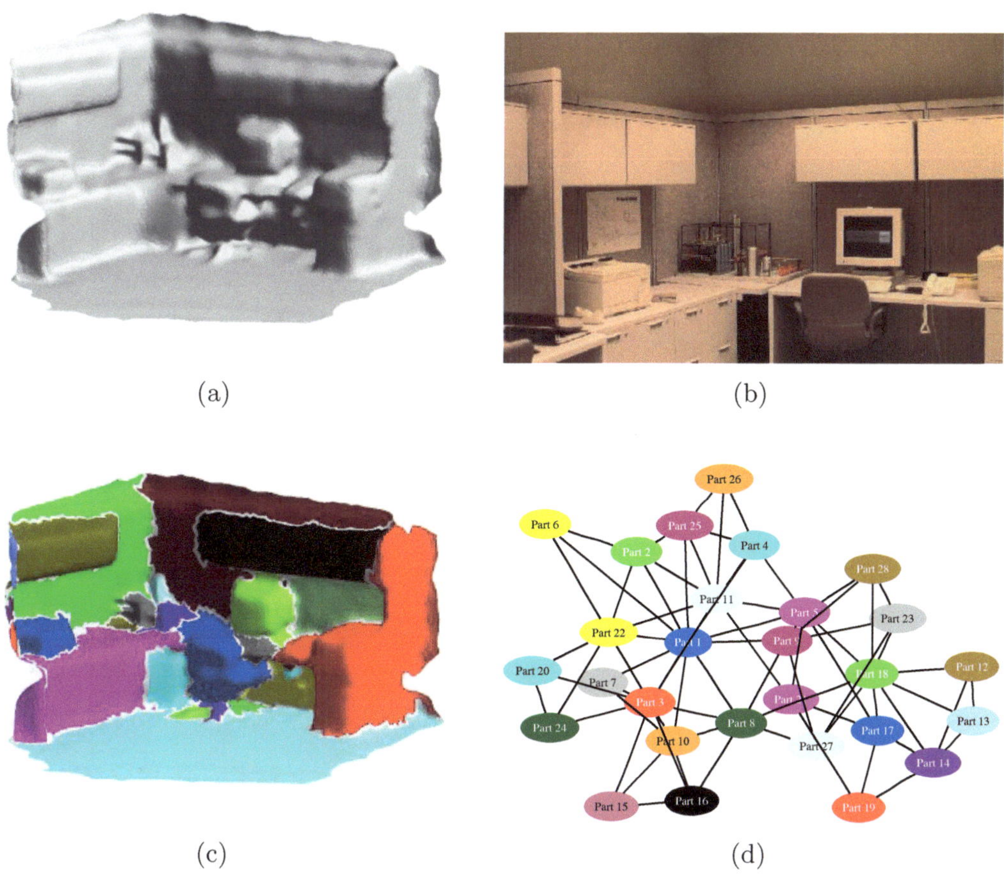

(a)

(b)

(c)

(d)

Figure 7.9: Minima Rule Algorithm results for an office scene. This mesh is a reconstruction from multiple range scans from the Perceptron Laser System. The mesh consists of $15,666$ vertices and $30,981$ triangles while the decomposition consists of 28 parts. (a) Rendering of original mesh. (b) Photograph of original scene. (c) Decomposition of mesh into parts. (d) Adjacency graph with arbitrary labels.

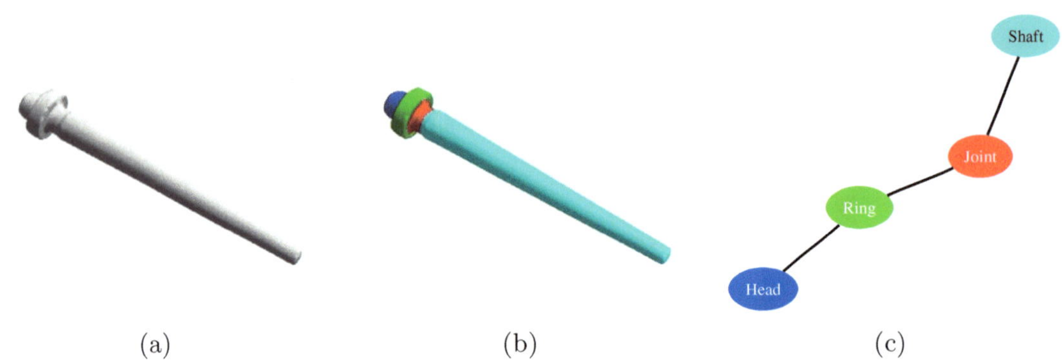

(a)   (b)   (c)

Figure 7.10: Minima Rule Algorithm results for a small bore pin. This mesh is a reconstruction from a point cloud data set generated at the U. S. Army TACOM National Automotive Center. The mesh consists of $37,4506$ vertices and $74,896$ triangles while the decomposition consists of 4 parts. (a) Rendering of original mesh. (b) Decomposition of mesh into parts. (c) Adjacency graph with user-specified labels.

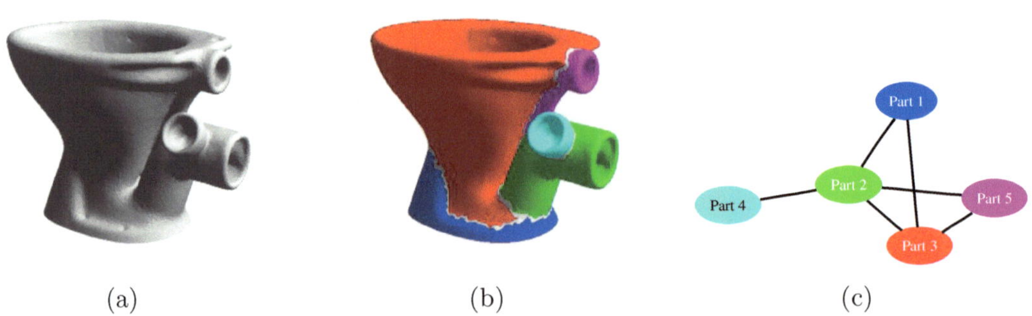

(a)   (b)   (c)

Figure 7.11: Minima Rule Algorithm results for a toilet seat. This mesh is a reconstruction from the Polhemus FastSCAN System. The mesh consists of $22,926$ vertices and $45,864$ triangles while the decomposition consists of 5 parts. (a) Rendering of original mesh. (b) Decomposition of mesh into parts. (c) Adjacency graph with arbitrary labels.

116

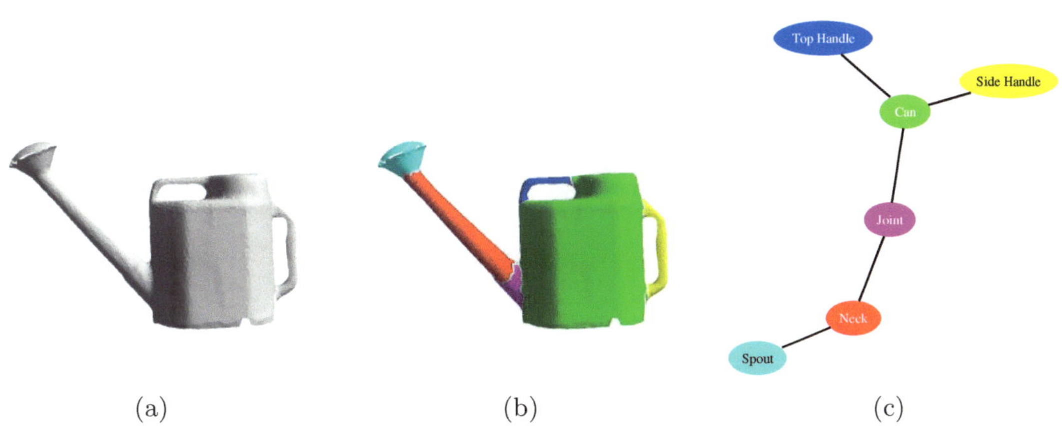

(a) (b) (c)

Figure 7.12: Minima Rule Algorithm results for a watering can. This mesh is a reconstruction from the Polhemus FastSCAN System. The mesh consists of $8,086$ vertices and $15,843$ triangles while the decomposition consists of 6 parts. (a) Rendering of original mesh. (b) Decomposition of mesh into parts. (c) Adjacency graph with user-specified labels.

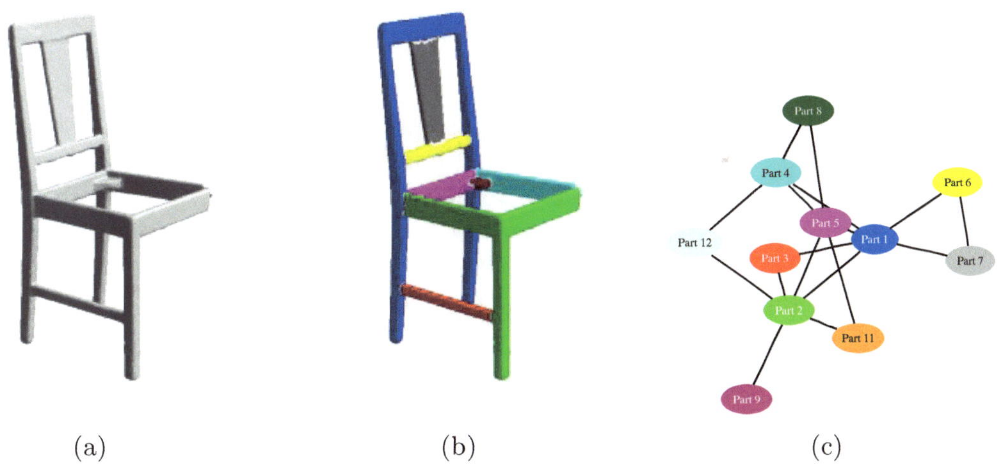

(a) (b) (c)

Figure 7.13: Minima Rule Algorithm results for a three-legged chair. This mesh is a reconstruction from the Polhemus FastSCAN System. The mesh consists of $26,766$ vertices and $53,462$ triangles while the decomposition consists of 11 parts. (a) Rendering of original mesh. (b) Decomposition of mesh into parts. (c) Adjacency graph with arbitrary labels.

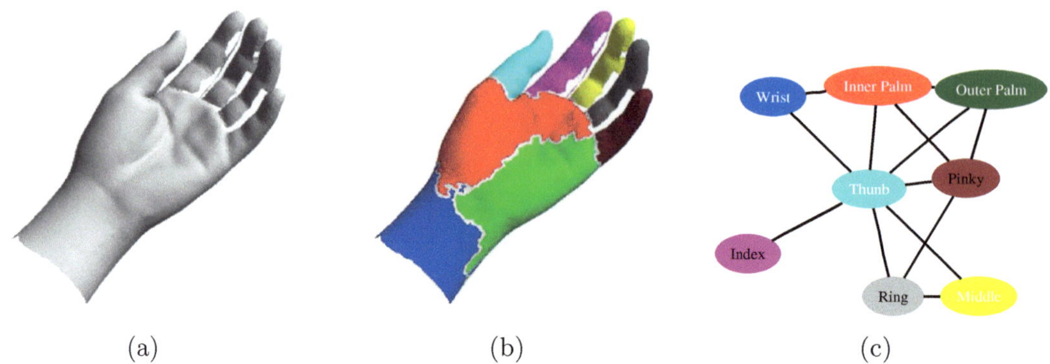

(a)   (b)   (c)

Figure 7.14: Minima Rule Algorithm results for a human left hand. This mesh is a reconstruction from the Polhemus FastSCAN System. The mesh consists of $13,340$ vertices and $26,372$ triangles while the decomposition consists of 8 parts. (a) Rendering of original mesh. (b) Decomposition of mesh into parts. (c) Adjacency graph with user-specified labels.

7.2.1 Qualitative Analysis

For a qualitative analysis, we first consider two illustrative data sets: a fandisk in Fig. 7.24 and a torus in Fig. 7.25. The fandisk is a CAD model that exemplifies free-form smooth surfaces, and the torus is a simple object with a genus one topology. We add noise to these data sets to demonstrate the robustness of our algorithm. To the original fandisk in Fig. 7.24(a), we have added Gaussian noise to each coordinate of the mesh's vertices where we specify the variance of the Gaussian distribution as 0.1% of the average edge length l_{ave} for triangles in the mesh. Fig. 7.24(b) illustrates this noise, which simulates the errors that we might encounter from reconstruction algorithms in computer vision. For the torus, we use a marching cubes algorithm (Lorenson and Cline, 1987) to introduce noise into the data set. Marching cubes is a common method in medical imaging to extract isosurfaces from volume data sets. The nature of the marching cubes algorithm corrupts the ideal smooth surface of the original torus in Fig. 7.25(a) with the more jagged surfaces in Fig. 7.25(b).

To begin, we explore the crease detection and normal estimation capabilities of our algorithm. Figs. 7.24(c) and 7.24(d) show the crease detection results for the fandisk both without noise and with it. Without noise, we design ε, which controls crease detection in Eq. (4.9), with a large enough value to detect accurately the crease on the left side of Fig. 7.24(c). With noise, however, we must decrease ε to account for the noise variation, and thus we no longer detect that particular crease in Fig. 7.24(d). We do, on the other hand, detect the right-angle creases for both cases of the fandisk despite the increase in noise. This example highlights the trade off between allowing noise variation and detecting surface discontinuities. When designing a system and thus choosing ε and η, the designer must balance the level of noise to tolerate and the angle of creases to detect. Then a user can choose various neighborhood sizes, as denoted by the

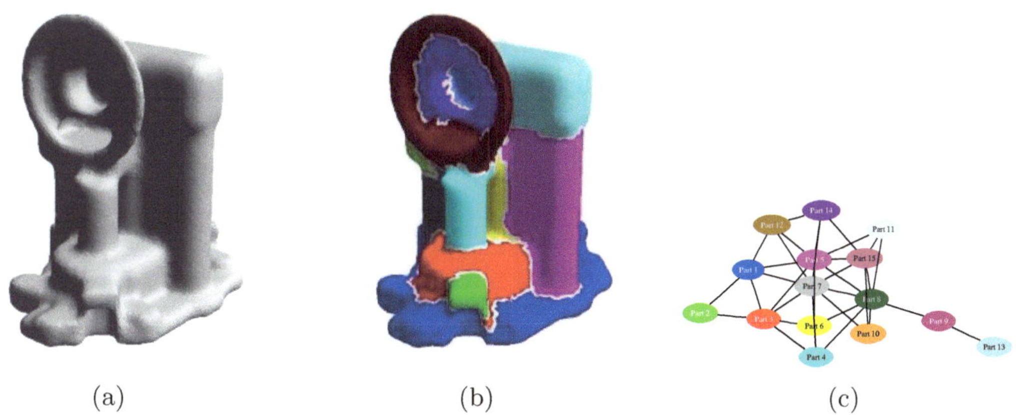

(a)   (b)   (c)

Figure 7.15: Minima Rule Algorithm results for an oil pump. This mesh is a reconstruction from the Hughes Hoppe at Microsoft Research. The mesh consists of $19,555$ vertices and $39,102$ triangles while the decomposition consists of 15 parts. (a) Rendering of original mesh. (b) Decomposition of mesh into parts. (c) Adjacency graph with arbitrary labels.

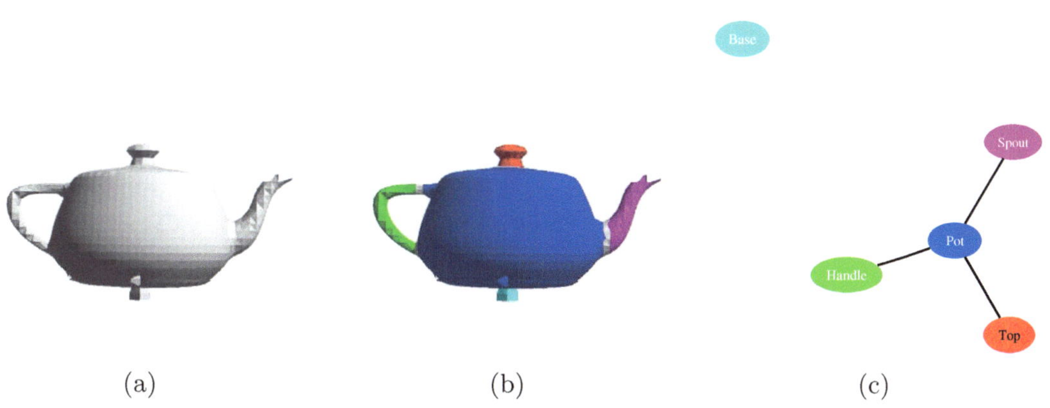

(a)   (b)   (c)

Figure 7.16: Minima Rule Algorithm results for a teapot. This mesh is a reconstruction from the Hughes Hoppe at Microsoft Research. The mesh consists of $3,034$ vertices and $6,010$ triangles while the decomposition consists of 5 parts. (a) Rendering of original mesh. (b) Decomposition of mesh into parts. (c) Adjacency graph with user-specified labels.

119

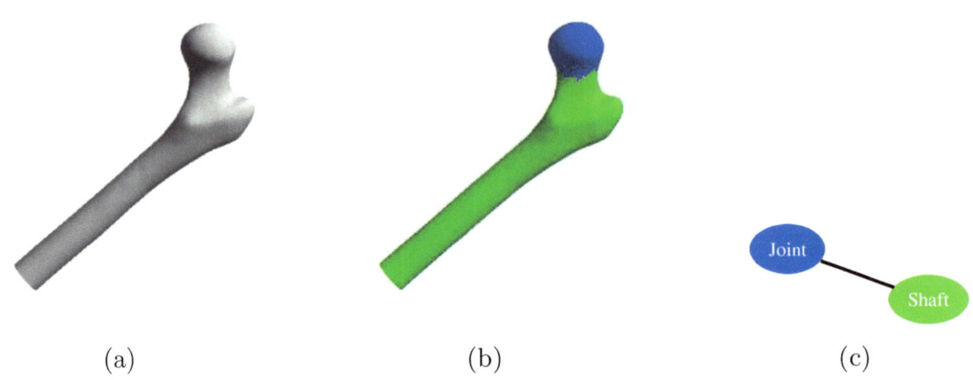

(a) (b) (c)

Figure 7.17: Minima Rule Algorithm results for a human femur. This finite element mesh has been created by the Marco Viceconti and is is available on the Internet at the ISB Finite Element Repository managed by the Istituti Ortopedici Rizzoli, Bologna, Italy. The mesh consists of $76,794$ vertices and $153,322$ triangles while the decomposition consists of 2 parts. (a) Rendering of original mesh. (b) Decomposition of mesh into parts. (c) Adjacency graph with user-specified labels.

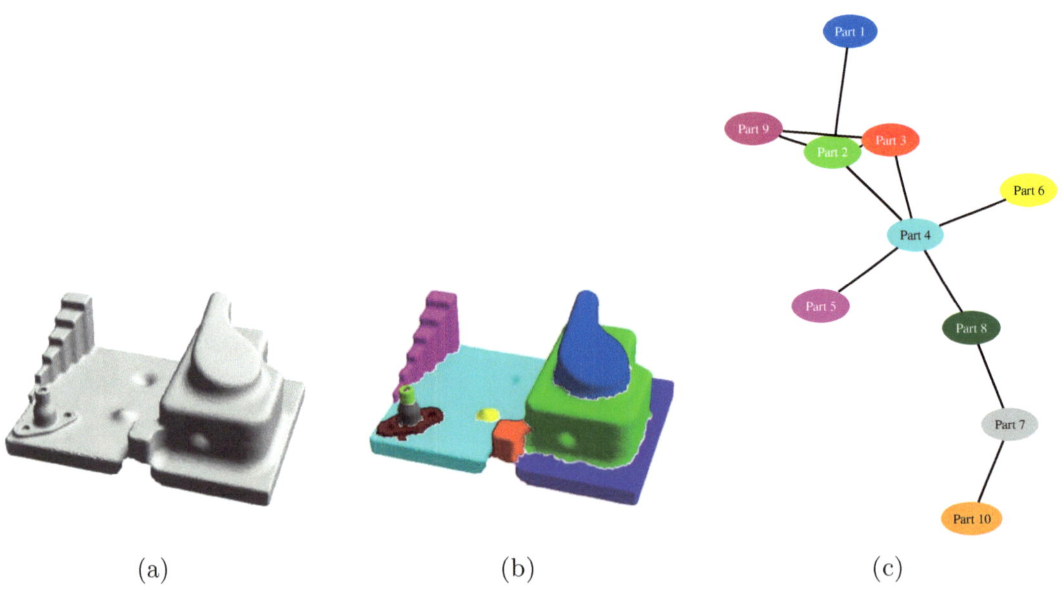

(a) (b) (c)

Figure 7.18: Minima Rule Algorithm results for a machined object. The mesh is a reconstruction using the SLIM3D software for reverse engineering. The mesh consists of $28,667$ vertices and $57,107$ triangles while the decomposition consists of 10 parts. (a) Rendering of original mesh. (b) Decomposition of mesh into parts. (c) Adjacency graph with arbitrary labels.

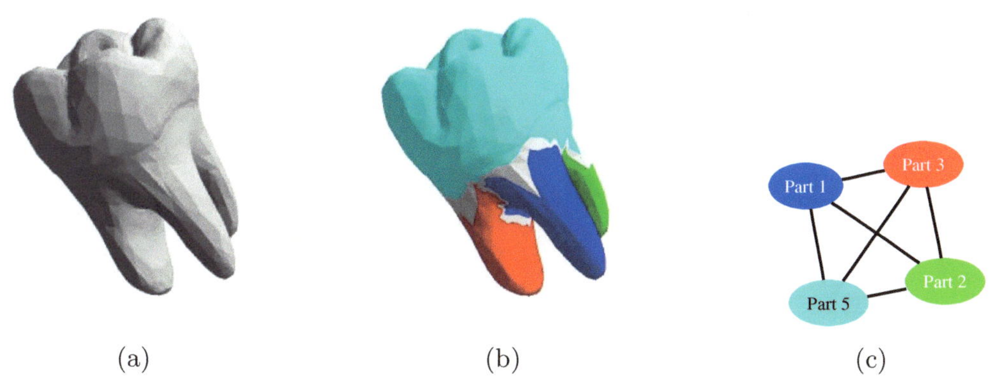

(a)   (b)   (c)

Figure 7.19: Minima Rule Algorithm results for a human molar tooth. The mesh is a reconstruction using the SLIM3D software for reverse engineering. The mesh consists of $6,586$ vertices and $13,168$ triangles while the decomposition consists of 4 parts. (a) Rendering of original mesh. (b) Decomposition of mesh into parts. (c) Adjacency graph with arbitrary labels.

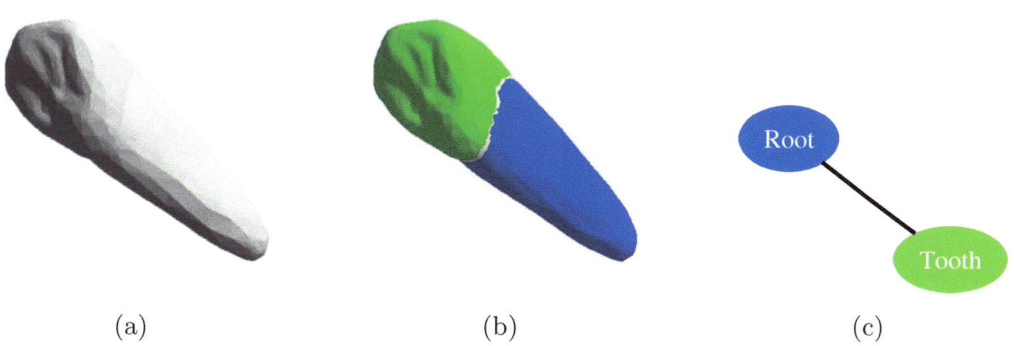

(a)   (b)   (c)

Figure 7.20: Minima Rule Algorithm results for a human canine tooth. The mesh is a reconstruction using the SLIM3D software for reverse engineering. The mesh consists of $3,376$ vertices and $6,748$ triangles while the decomposition consists of 2 parts. (a) Rendering of original mesh. (b) Decomposition of mesh into parts. (c) Adjacency graph with user-specified labels.

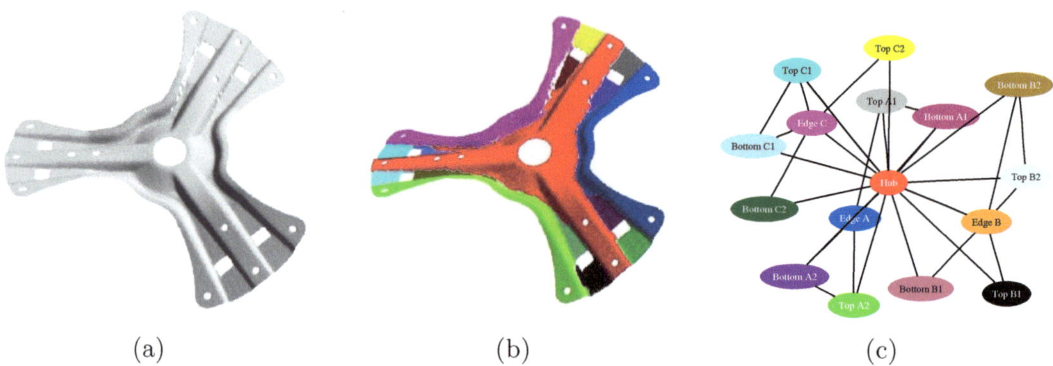

(a)   (b)   (c)

Figure 7.21: Minima Rule Algorithm results for a industrial fan. The mesh is a reconstruction from the 3D Digital Corporation. The mesh consists of $121,271$ vertices and $239,227$ triangles while the decomposition consists of 16 parts. (a) Rendering of original mesh. (b) Decomposition of mesh into parts. (c) Adjacency graph with user-specified labels.

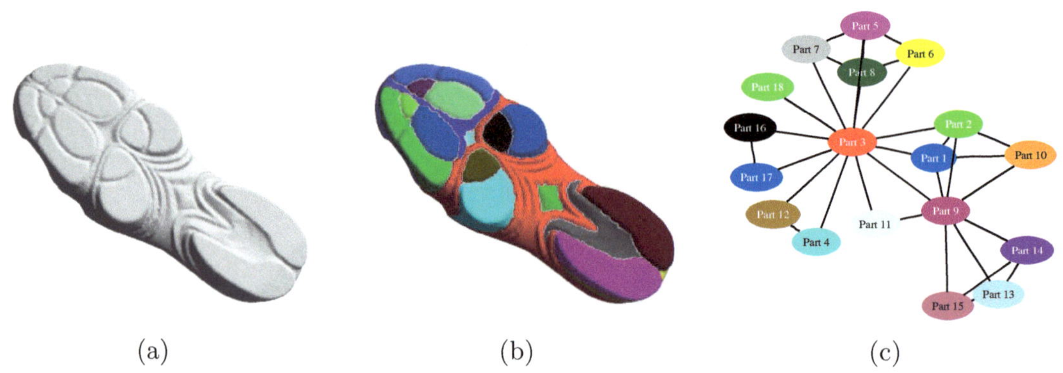

(a)   (b)   (c)

Figure 7.22: Minima Rule Algorithm results for a shoe sole. The mesh is a reconstruction from the 3D Digital Corporation. The mesh consists of $58,108$ vertices and $115,750$ triangles while the decomposition consists of 18 parts. (a) Rendering of original mesh. (b) Decomposition of mesh into parts. (c) Adjacency graph with arbitrary labels.

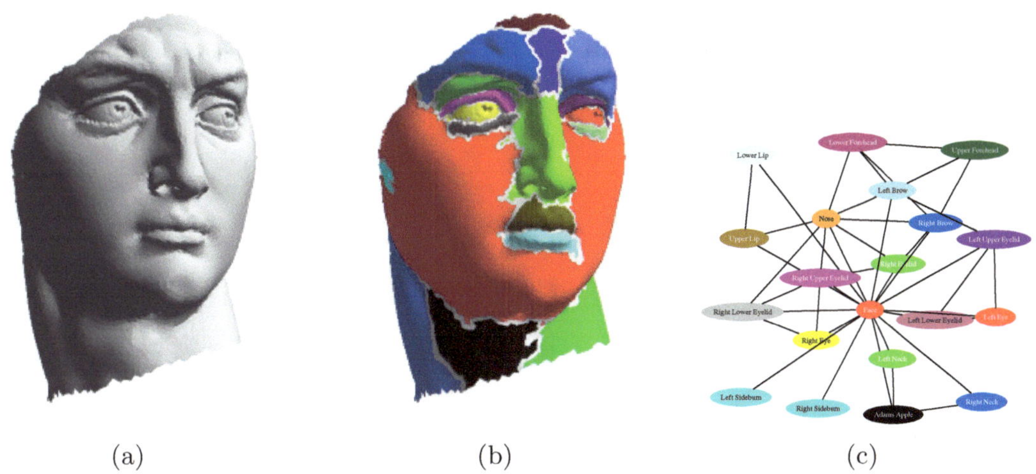

(a)   (b)   (c)

Figure 7.23: Minima Rule Algorithm results for the head of Michelangelo's David statue. The mesh is a reconstruction from The Digital Michelangelo Project at Stanford University. We have extracted just the face from a more complete model of the entire statue. The mesh consists of $7,790$ vertices and $15,203$ triangles while the decomposition consists of 20 parts. (a) Rendering of original mesh. (b) Decomposition of mesh into parts. (c) Adjacency graph with user-specified labels.

user parameter k from Table 4.2, to balance noise levels, feature scales, and sampling densities for a particular mesh.

Next, the torus example illustrates normal estimation. Fig. 7.25(c) shows a zoom view for the edge of the torus. In this figure, the vectors extending from the mesh's vertices are normal estimates derived from a simple average of the triangle normals adjacent to each vertex. Fig. 7.25(d) shows the same view but with normals derived from the Normal Voting algorithm. The neighborhood size for the latter is an eight-geodesic neighborhood ($k = 8$). The larger neighborhood allows for an improvement in the estimation of the normals. We see that the normals in Fig. 7.25(c) adhere to the marching cube artifacts while in Fig. 7.25(d) they more closely follow the smooth surface of the original torus. This smoothing characteristic of a larger neighborhood is the primary strength of the Normal Vector Voting algorithm.

The third example is in Figure 7.26(a). This figure is a 3D mesh from a Perceptron laser range scanner and is representative of the surface noise that one might expect from a practical system. We apply the normal voting algorithm to this mesh with the results in Figure 7.26(b). We use a $k = 5$ voting neighborhood, and we place blue normals at surface vertices and green tangents at crease ones. Additionally, we place red vectors at vertices with no preferred orientation. The algorithm performs well for detecting the creases between the floor and the base of each desk and around the cabinets, chair, and computer monitor in the scene. Unfortunately, the corner of the room results in several vertices classified as having no preferred orientation. In the actual scene, this corner contains an intricate scaled model of an industrial plant with several small

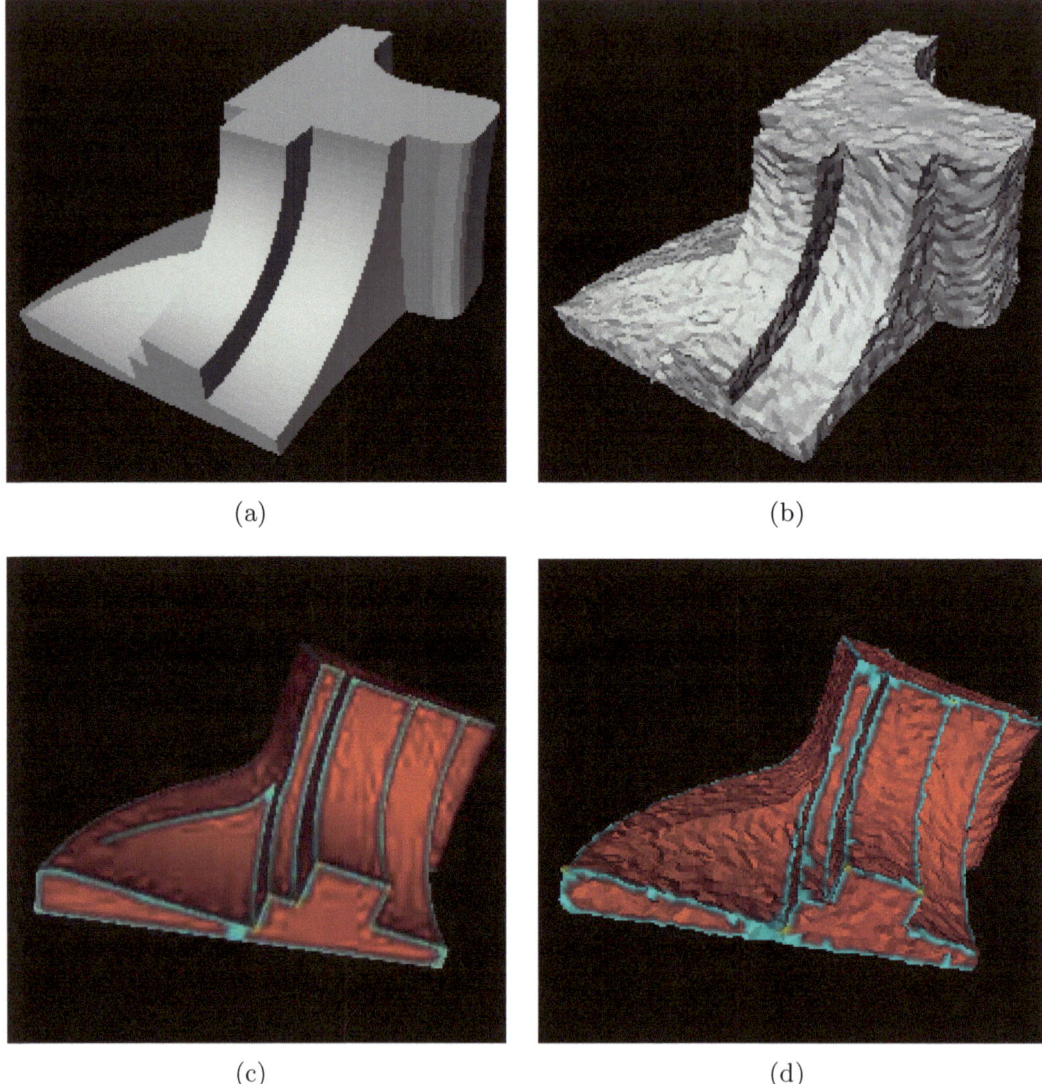

(a)　　　　　　　　　　　　(b)

(c)　　　　　　　　　　　　(d)

Figure 7.24: The fandisk model is an example of a free-form CAD model with sharp edges and sophisticated surface curvature. (a) The original triangle mesh is from the distribution available at `http://research.microsoft.com/research/graphics/hoppe/`. The mesh size is 6,475 vertices with 12,946 triangles. (b) Same model with Gaussian noise added to the vertices of the mesh. The variance of the Gaussian distribution is 0.1% of l_{ave}. (c) The Normal Vector Voting algorithm with a one-geodesic neighborhood ($\varepsilon = 32$, $k = 1$) labels the creases of the original mesh as we might expect. (d) With noise the algorithm using a five-geodesic neighborhood ($\varepsilon = 2$, $k = 5$) does not detect the small crease on the left side of the figure.

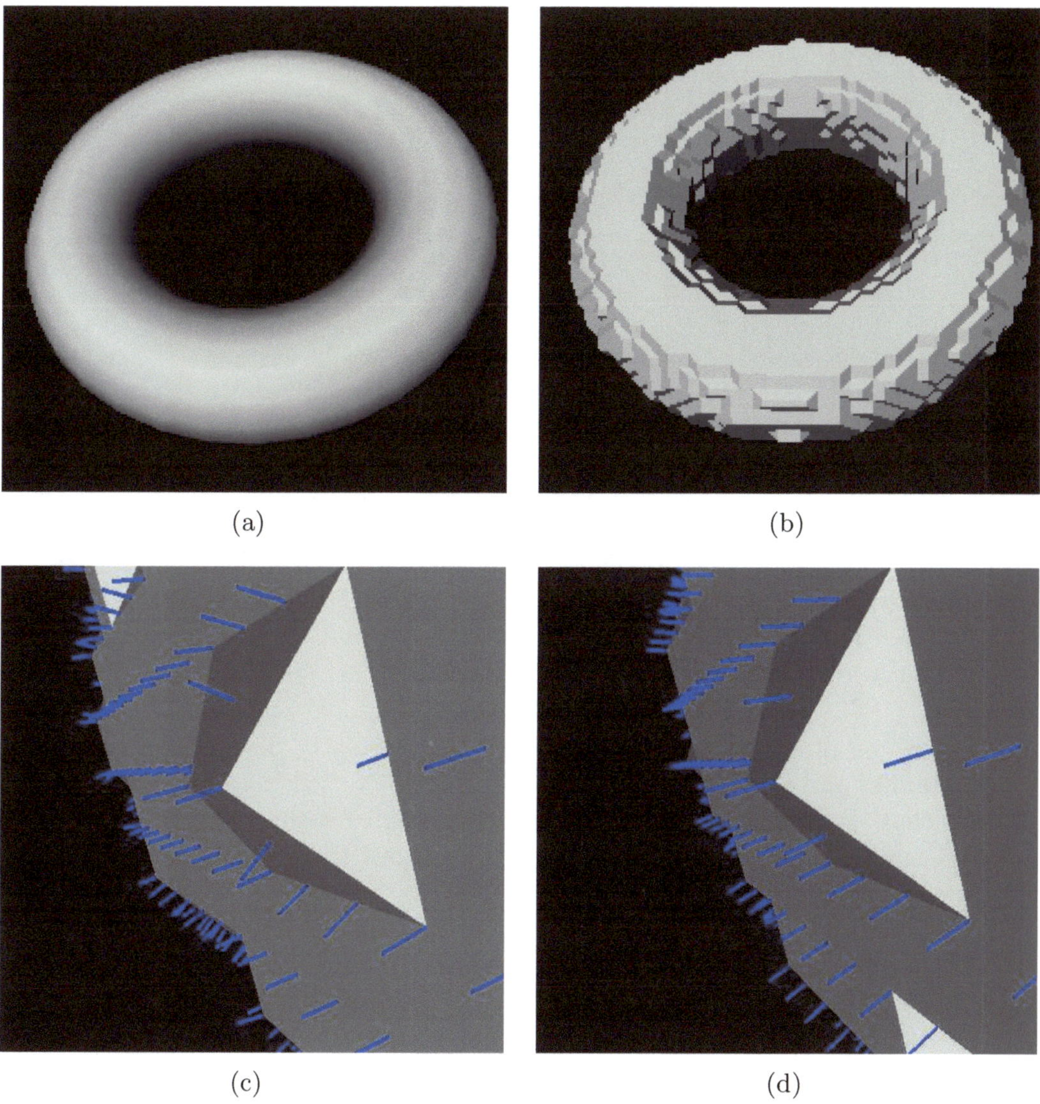

(a)

(b)

(c)

(d)

Figure 7.25: The torus model exhibits a genus one topology. (a) The smooth surfaces of the original model. (b) If we consider the torus as a volume data set and apply a marching cube algorithm, we introduce systematic artifacts in the mesh. This mesh has 7,302 vertices and 14,604 triangles. (c) A zoom view of the side of the mesh in (b) with normals shown extending from each vertex. (d) The same zoom area where the normals are estimated using Normal Vector Voting with an eight-geodesic neighborhood ($k = 8$).

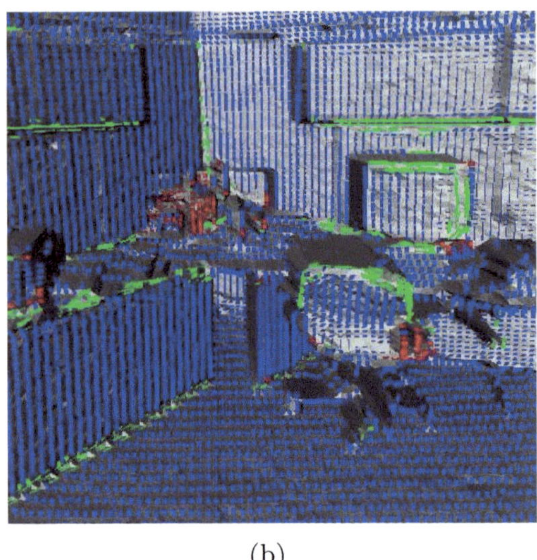

(a)   (b)

Figure 7.26: Data from a Perceptron range scanner. (a) A triangle mesh with 16,384 vertices and 32,258 triangles for the corner of a room, which contains a desk, a chair, a computer monitor, and a few cabinets. The surface noise is both quantization and measurement error inherent to the scanner. (b) Normal Vector Voting leads to blue normals at each surface patch vertex, green tangents at each crease vertex, and red vectors at each vertex with no preferred orientation.

wooden dowels as pipe models. The detail of this model is beyond the resolution of the range scanner and therefore the classification is not surprising. The conclusion that we draw is that the resolution of the mesh is too small for the scene features and not that the normal voting algorithm has failed. We suggest that the ability to detect such under-sampled regions is an important capability of our algorithm.

Figs. 7.27 through 7.29 demonstrate the robustness of the curvature estimate. Again, we use the fandisk and torus for these examples. In these figures, the vector at each vertex is the estimate for the principal direction of the maximum curvature. Figs. 7.27 and 7.28 show the effects of the neighborhood size. The small neighborhoods in Figs. 7.27(c) and 7.28(c) are unable to overcome the surface noise such that the zoom views reveal the random distribution of the principal directions. The larger neighborhoods in Figs. 7.27(d) and 7.28(d) yield improvement with strong local agreement of the principal directions shown in the zoom views. Notice how the principal directions flow along the surface curvature of the object. These figures highlight the robustness of the curvature estimation via selection of the neighborhood size. As noted previously, this capability is the major contribution of this paper.

One drawback to enlarging the neighborhood is that significant error is introduced at discontinuity creases. Consider one of the creases in the fandisk model. Fig. 7.29(a) shows a crease at the top of the figure with corresponding crease tangents. The principal directions just below that crease erroneously point across the crease. The problem is

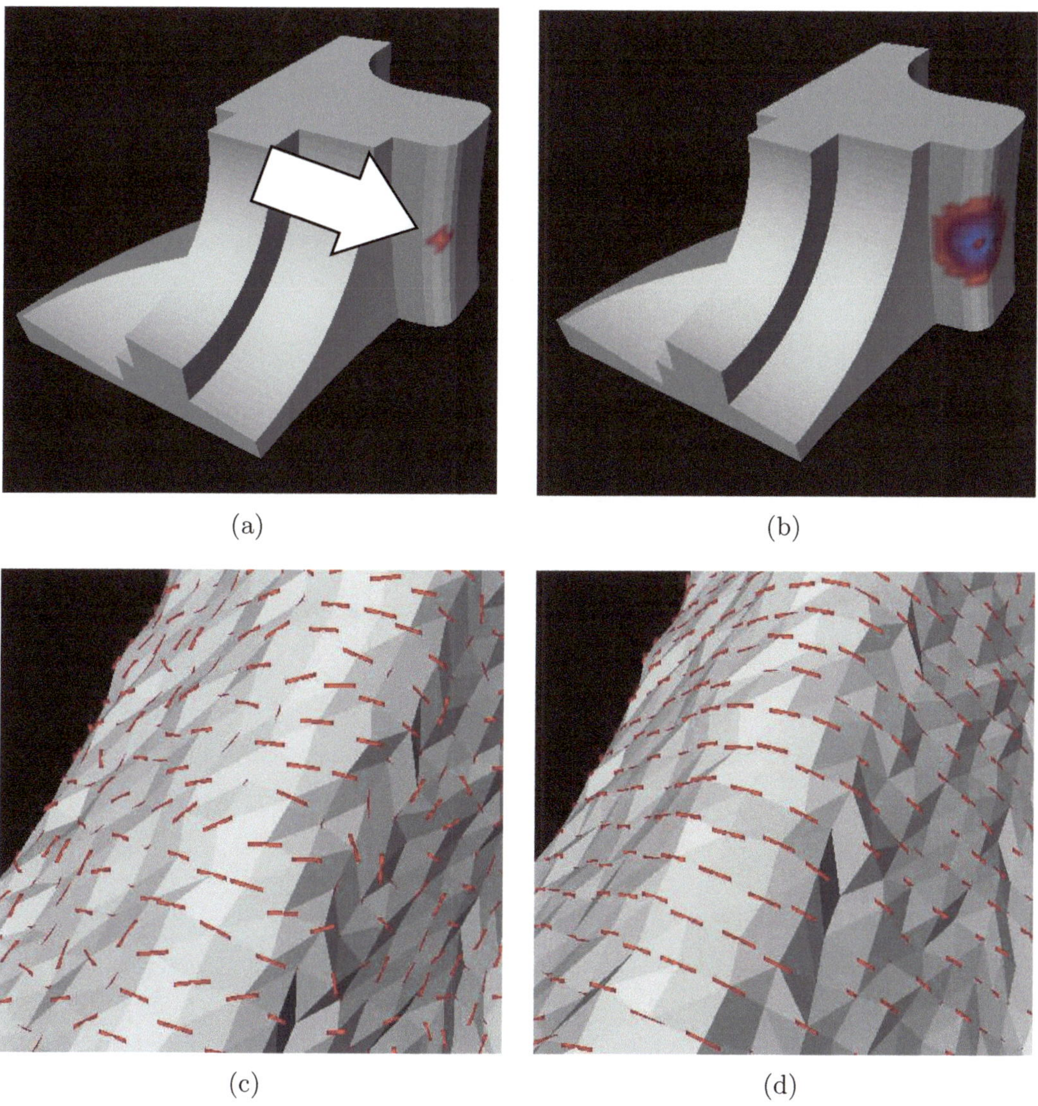

Figure 7.27: Curvature estimation for the fandisk model. (a) Illustrates a small one-geodesic neighborhood ($k = 1$) for a single vertex. (b) A larger five-geodesic neighborhood ($k = 5$) for the same vertex. (c) Estimates for the principal directions with 0.1% Gaussian noise added to the fandisk. The zoom area is for the small neighborhood in (a). (d) The same zoom area showing estimates for the larger neighborhood of (b).

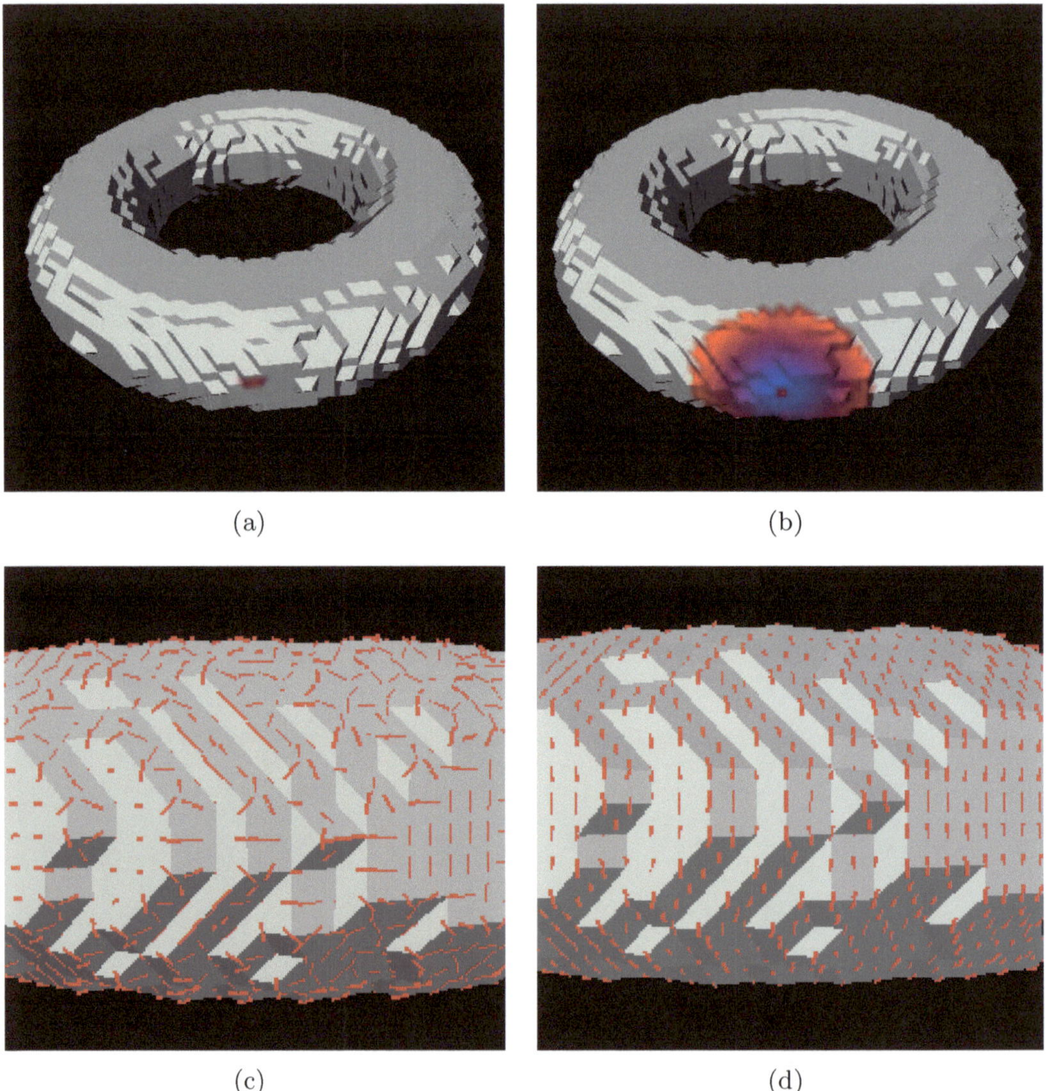

(a)

(b)

(c)

(d)

Figure 7.28: Curvature estimation for the torus. (a) Illustrates a small one-geodesic neighborhood ($k = 1$) for a single vertex. (b) A larger eight-geodesic neighborhood ($k = 8$) for the same vertex. (c) Shows the principal directions for small neighborhood in (a). (d) The same zoom area showing estimates for the larger neighborhood of (b).

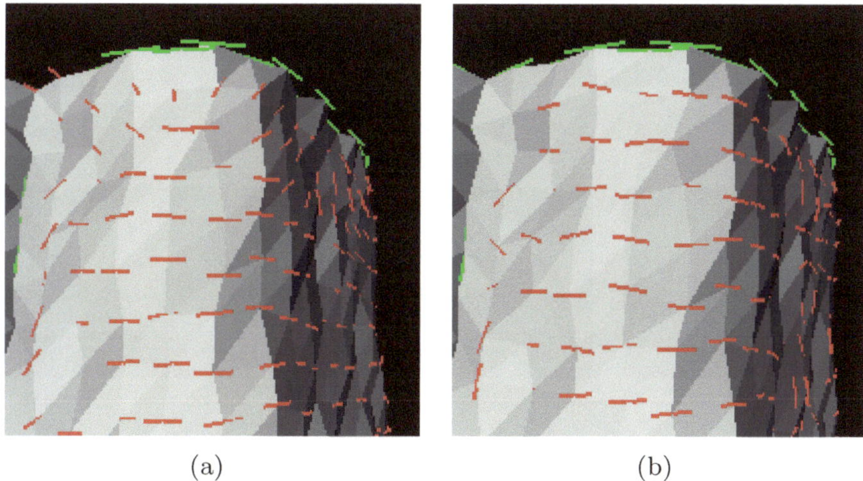

(a)   (b)

Figure 7.29: Increasing the neighborhood leads to errors in the curvature estimate near the crease discontinuities. (a) The zoom view shows a crease from the fandisk at the top of this image. The principal directions near this crease tend to point erroneously across the discontinuity. (b) When we restrict the neighborhood from crossing the discontinuity, the estimates for the principal directions improve.

that normals on the other side of the discontinuity vote during the curvature estimation. To resolve this problem, we restrict the neighborhood from crossing the discontinuity. Fig. 7.29(b) illustrates the results. The principal directions near the crease now follow the curvature of the surface as we desire.

7.2.2 Quantitative Comparison

The previous examples show the capabilities of the Normal Vector Voting algorithm but do not provide a baseline for comparison. Figs. 7.31, 7.32, and 7.35 attempt to do so. In these figures, we graph the error of the algorithm for both synthetic and real data where we use ground truth to establish the error. For these graphs, we manipulate three variables: surface type, noise level, and neighborhood size.

We first consider synthetic data to evaluate noise sensitivity. We choose three surface types: planar, cylindrical, and spherical. The radius of curvature for the cylinder and the sphere is 31 mm. Fig. 7.30 shows these surfaces with Gaussian noise added to the Z coordinate using a distribution variance of 1%, 5%, 10%, and 50% of l_{ave}. The graphs in Fig. 7.31 compare the performance of the Normal Vector Voting algorithm for a three- and five-geodesic neighborhood ($k = 3$ and $k = 5$) to Taubin's one-ring algorithm for 10% noise level. We have chosen to compare to Taubin's algorithm since it serves as the foundation for our algorithm. We do not compare results for Tang and Medioni's algorithm since their algorithm is for point-cloud data sets and since they do not compute principal directions for surface curvature.

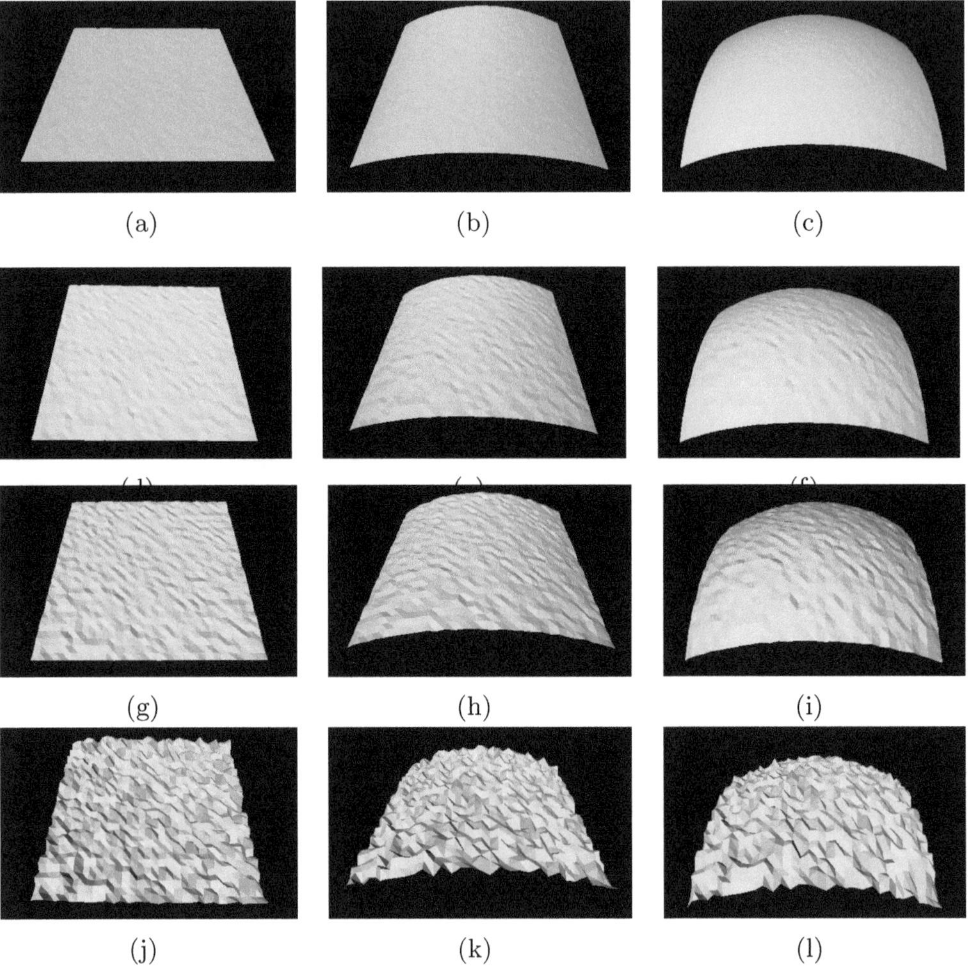

Figure 7.30: Planar, cylindrical, and spherical synthetic surfaces with four levels of Gaussian noise. These surfaces contain 1,024 vertices and 1,922 triangles. (a–c) Noise variance is 1% of the average edge length. (d–f) Noise variance is 5% of the average edge length. (g–i) Noise variance is 10% of the average edge length. (j–l) Noise variance is 50% of the average edge length.

Taubin's algorithm serves to baseline the performance of our method. However, for noisy surfaces, a direct comparison is perhaps not appropriate since—as Taubin notes—his algorithm requires a smoothing step to preprocess any surface noise. In fairness to Taubin, we emphasize that we baseline the benefits of the k-geodesic neighbors and do not competitively compare the algorithms. To do so, we would need to select and implement an appropriate smoothing algorithm.

We look at the normal estimation for the plane in Fig. 7.31(a). The percentage error in this graph is as follows

$$err = (1 - |N_p^t N_v|) \times 100\% \tag{7.1}$$

where $N_p = Z$ is the ground-truth normal and N_v is the estimation. This graph is a histogram plot with vertex bins across the horizontal axis and a log scale for the vertical. Fig. 7.31(b) uses a similar error measure and compares the estimation of the principal directions T_v for the cylinder. Let $T_p = X$ be the ground truth for the minimum principal direction. The third graph in Fig. 7.31(c) compares the estimation of the principal curvatures for the sphere. We use a different error measure

$$err = \left| \frac{\kappa_p^1 - \kappa_v^1}{\kappa_v^1} \right| \times 100\% \tag{7.2}$$

where $\kappa_p^1 = \frac{1}{31mm}$ is the ground truth and κ_v^1 is the estimate. For each of these graphs, we see a similar trend. The Normal Vector Voting algorithm for both neighborhoods provides improved performance over Taubin's algorithm. This improvement is evident for bins near the 0% error to the left side of the graphs. The more vertices that accumulate in these lower error bins the better.

We next consider the effect of different noise levels. As before for each synthetic surface, we corrupt the Z coordinate of each vertex with Gaussian noise. Recall Fig. 7.30. Using the previous error measures and graphs, Figs. 7.32 plots the Normal Vector Voting error for surface type and noise level with only the three-geodesic neighborhood size ($k = 3$). Although the 50% level seems to overwhelm the Normal Vector Voting algorithm, the other three levels offer useful results for most applications. Again, these graphs demonstrate the robustness of the algorithm.

Finally, we explore real data from an IVP Ranger System (Integrated Vision Products, 2000), which is a sheet-of-light profile scanner. Fig. 7.33(a) shows the experimental configuration for this system within our laboratory. The basic output of the scanner is a single range profile in the plane of the sheet of light. For our tests we stack 512 profiles together to form a 512×512 range image with 256 range bins at 0.62 mm resolution. With proper system calibration, we convert these range images into appropriate triangle meshes. We again use three surface types as with the synthetic data. The actual objects for these surfaces appear in Fig. 7.33(b). As ground truth, the cylinder has a radius of 26 mm, and the sphere has a radius of 28.5 mm. With slight modifications, we use the same error measures and graphs as with the synthetic data. Since we do not know the absolute orientation of the objects relative to the scanner, we must account

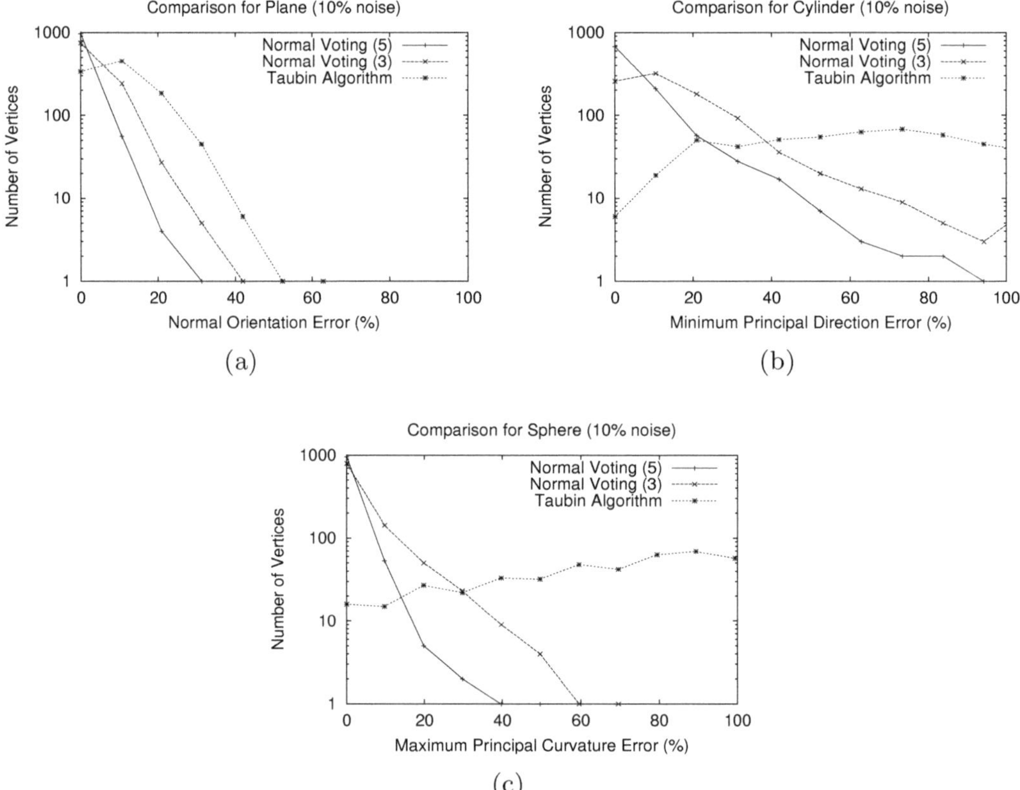

Figure 7.31: Algorithm comparison for synthetic surfaces with Gaussian noise added. The variance of the noise distribution is 10% of l_{ave}. The comparison graphs show the Normal Vector Voting algorithm with a three- and a five-geodesic neighborhood size ($k = 3$ and $k = 5$) compared to the original algorithm of Taubin. The graphs are bin plots with log scales, and each mesh contains 1,024 vertices and 1,922 triangles. (a,b) Planar surface (c,d) Cylindrical surface, and (e,f) Spherical surface.

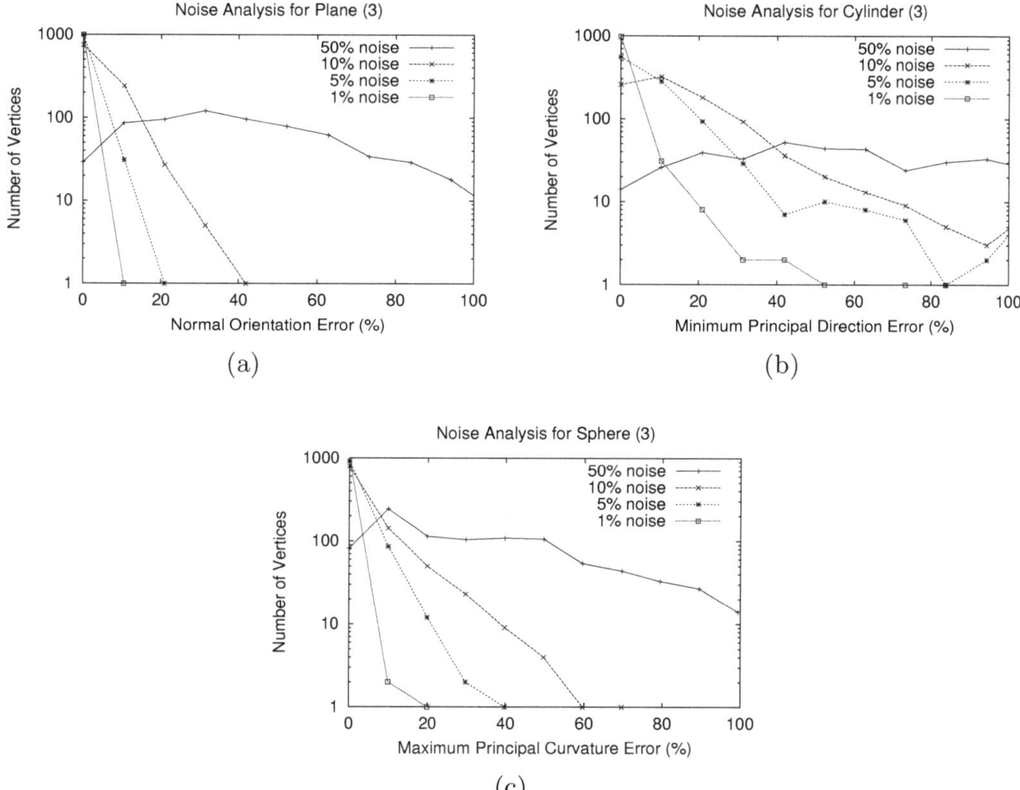

Figure 7.32: Comparison for synthetic data with different levels of Gaussian noise added. The variances of the noise distribution are 1%, 5%, 10%, and 50% of l_{ave}. The graphs show the Normal Vector Voting algorithm for a three-geodesic neighborhood ($k = 3$) and are bin plots with log scales. Each mesh contains 1,024 vertices and 1,922 triangles. (a,b) Plane, (c,d) Cylinder, and (e,f) Sphere.

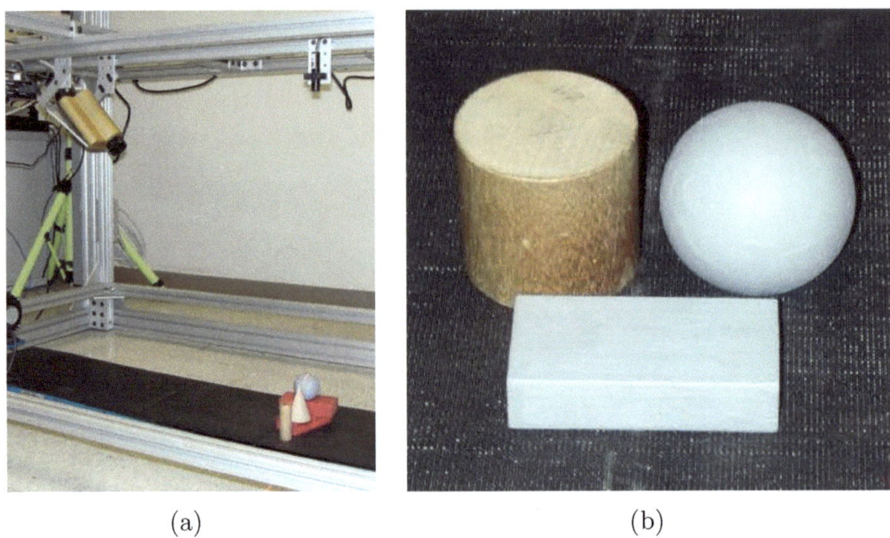

(a)   (b)

Figure 7.33: IVP Ranger System, which is a sheet-of-light profiling system. (a) Experimental setup. (b) Ground truth objects for our real range data comparison.

for this uncertainty. For the plane, we average the normal estimates to serve as the ground truth $N_p = \frac{1}{n} \sum N_v$ for each vertex, and for the cylinder, we average the minimum principal direction estimates $T_p = \frac{1}{n} \sum T_v$. These *ad hoc* formulations of ground truth introduce some error, but the error is constant across our analysis and is tolerable. Fig. 7.35 shows the scanned surfaces one the left column and the error graphs on the right. These results are similar to the synthetic data where again the Normal Vector Voting algorithm shows improvement over Taubin's algorithm.

7.2.3 Timing Performance

An issue of future research is to improve the overall performance of our implementation. The graph in Fig. 7.36 shows a simple timing analysis of the proposed algorithm for the fandisk model in Fig. 7.24. The computing platform is an SGI Octane with a single 195 Mhz MIPS R10000 processor and 128 megabytes of memory. Although we have not optimized the current code configuration, the trend in the plot is interesting. As we increase the k-geodesic neighborhood with the intent of improving our curvature estimate, the computational time grows non-linearly. The three- and five-geodesic neighborhoods show reasonable performance compared to the one-geodesic neighborhood that is equivalent to the one-ring algorithm of Taubin. Thus we argue that the improvement in accuracy, as demonstrated in Figs. 7.31–7.35, with modest increases in neighborhood size are worth the slight increase in processing time.

We have presented results for our algorithm using both synthetic and real data sets. For the synthetic data, we have used controlled experiments with ground truth to evaluate the performance of the algorithm with respect to varying levels of noise. For the

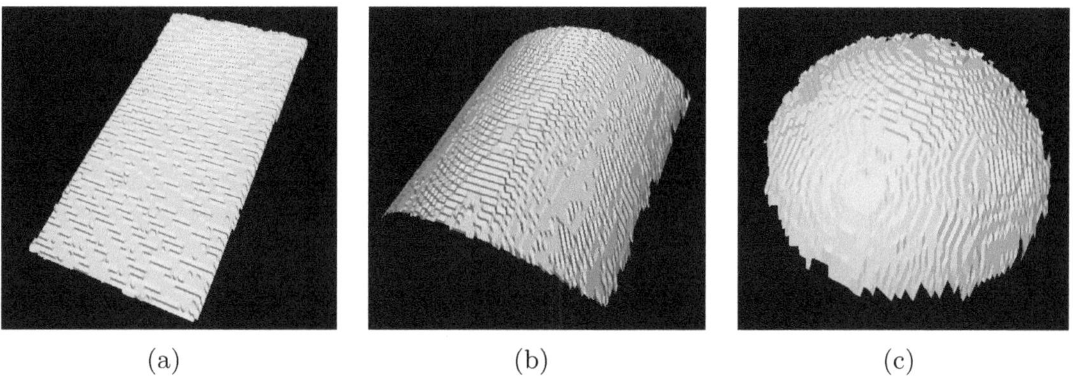

(a)   (b)   (c)

Figure 7.34: Triangle mesh representations of a single view for each ground truth object. The noise on the surfaces is actual noise from either quantization error or direct measurement error. (a) Plane tilted at 45 degree angle to viewing plane of Ranger. (b) Clynder parallel to viewing plane. (c) Sphere.

real data, we have generated meshes within our laboratory using the IVP Ranger. In addition to a qualitative analysis of our results, we have presented an in-depth quantitative analysis as well. In particular, we have directly compared the results of our algorithm to Taubin's original algorithm (Taubin, 1995). As a side note, since Tang and Medioni (Tang and Medioni, 1999) formulate their algorithm for point clouds and since they do not estimate principal curvatures, we do not compare our algorithm to theirs. The success of these results demonstrate the stable and robust performance of our algorithm in the presence of different types of surface noise. In the next section, we explore results for the Fast Marching Watersheds algorithm which leverages the curvature estimation results from this section.

7.3 Fast Marching Watersheds

After we have estimates for the principal curvatures and their associated directions for each vertex of the mesh, we need an algorithm that segments the mesh and identifies contours of negative curvature minima. The algorithm that we have developed for this purpose is the Fast Marching Watersheds algorithm. In this section, we analyze this algorithm through experimental results to find its strengths and weaknesses. Further, we investigate the effects of user parameters on the performance of the algorithm. Recall the parameters in Table 5.2. The two parameters are: α, the percentage of threshold for the marker set, and k, the ring size of the D_k structuring element. We mainly control the algorithm with α. The familiar mug example in Fig. 7.37 shows the progression of the algorithm.

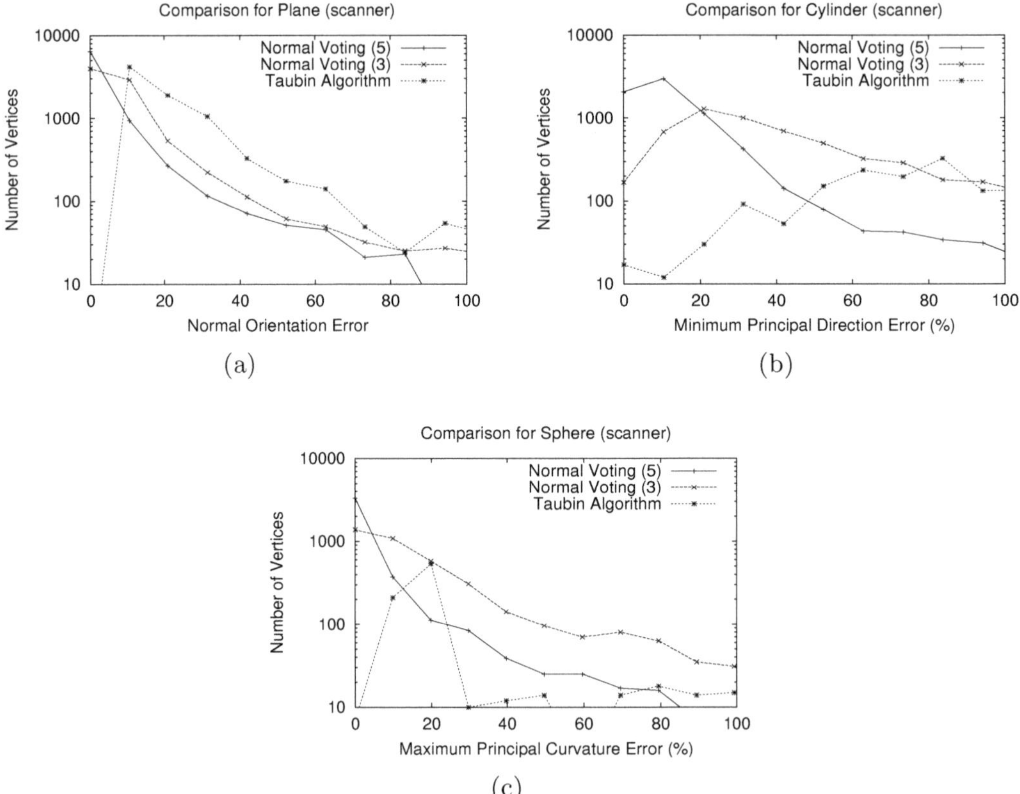

Figure 7.35: Algorithm comparison for real data from IVP Ranger. The comparison graphs show the Normal Vector Voting algorithm with a three- and a five-geodesic neighborhood size ($k = 3$ and $k = 5$) compared to the original algorithm of Taubin. The graphs are bin plots with log scales. Note the quantization noise on each surface. (a,b) Plane with 10,507 vertices and 20,578 triangles. (c,d) Cylinder with 7,407 vertices and 14,424 triangles. (e,f) Sphere with 4,026 vertices and 7,732 triangles.

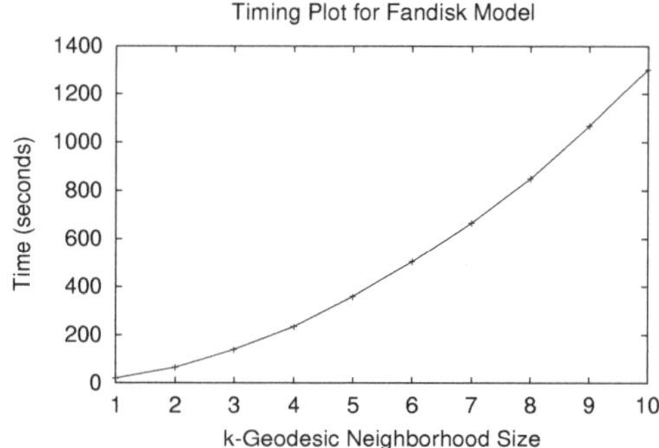

Figure 7.36: A graph depicting the computational performance time for the Normal Vector Voting algorithm as a function of k-geodesic neighborhood size. The analysis is for the fandisk model and the times are in seconds. Each data point on the graph represents a single run of the two-pass algorithm through the entire mesh.

7.3.1 Qualitative Comparison

We begin with a qualitative analysis. We first compare the algorithm to others in the literature, and then we investigate a potential drawback of the algorithm.

The examples in Figs. 7.38 and 7.39 serve to compare our segmentation to two algorithms in the literature. The first one is Mangan and Whitaker (Mangan and Whitaker, 1999), and the second is a variation of Pulla (Pulla, 2001). We do not directly test these algorithms since the source code is unavailable. We could have implemented our own code based on the papers, but such an approach does not always replicate the nuances derived from the experience of the paper's authors. An in-house implementation without source code is unreliable and is not fair for comparison purposes. However, we are able to implement their definitions of curvature reliably and without ambiguity. So, for the comparisons that follow, we have implemented their definitions of curvature and applied our watershed algorithm—with slight modifications—to achieve the spirit of the results in (Mangan and Whitaker, 1999) and (Pulla, 2001). We note that this approach does not allow a quantitative comparison of the algorithms, but it does allow a qualitative comparison of expected results.

The first scene in Fig. 7.38 for our comparison is a synthetic mesh representing a cylindrical barrel resting on a floor. The surface mesh approximates both the barrel and the floor with $5,313$ vertices and $10,374$ triangles. We first apply Mangan and Whitaker's segmentation in Fig. 7.38(b). Their algorithm generates three regions: the floor, the barrel side, and the barrel top. Their method segments regions bounded by contours of high curvature without regard to the positive or negative sense. In fact, Mangan and Whitaker's curvature measure exhibits no sign information and is always positive. Consequently, their segmentations are surfaces with similar values of curvature and do not necessarily agree with human perceptional segmentations. Most human observers would segment the scene into only two parts: the floor and the barrel without distinguishing the top from the sides of the barrel.

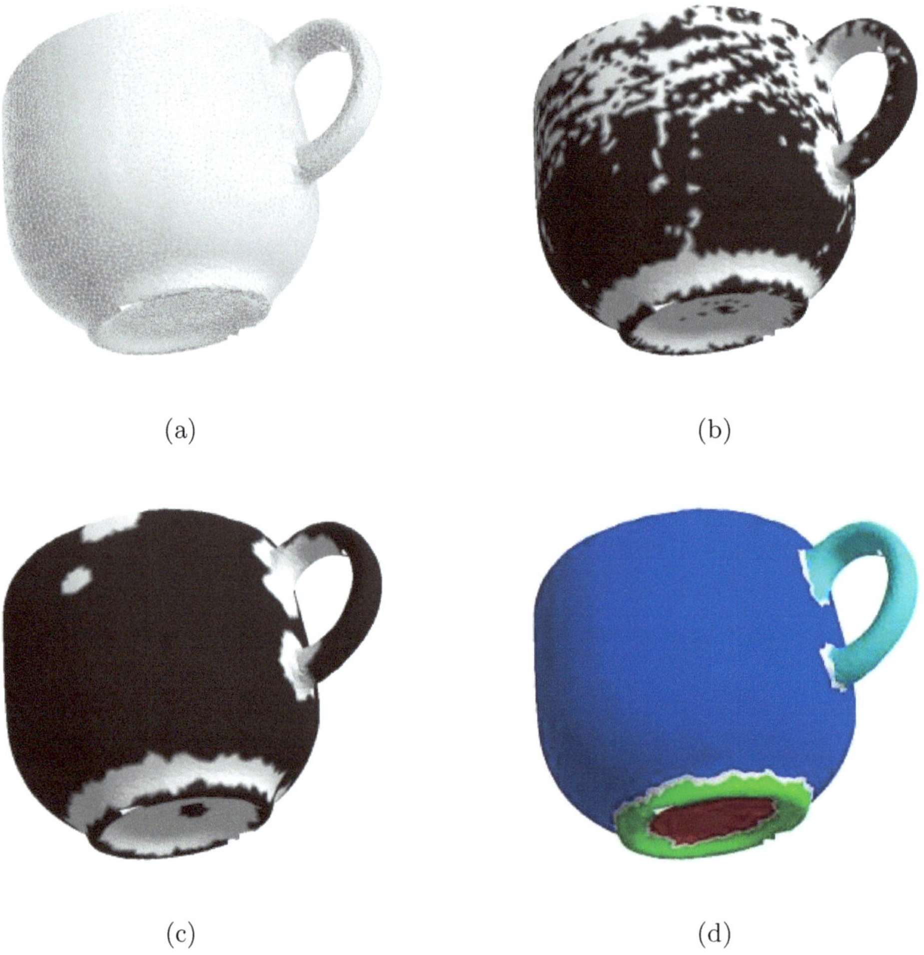

(a)

(b)

(c)

(d)

Figure 7.37: This example shows the progression of Fast Marching Watersheds. (a) The familiar coffee mug. (b) An initial threshold of curvature. (c) Morphology operators clean the marker set. (d) Final segmented regions after fast marching.

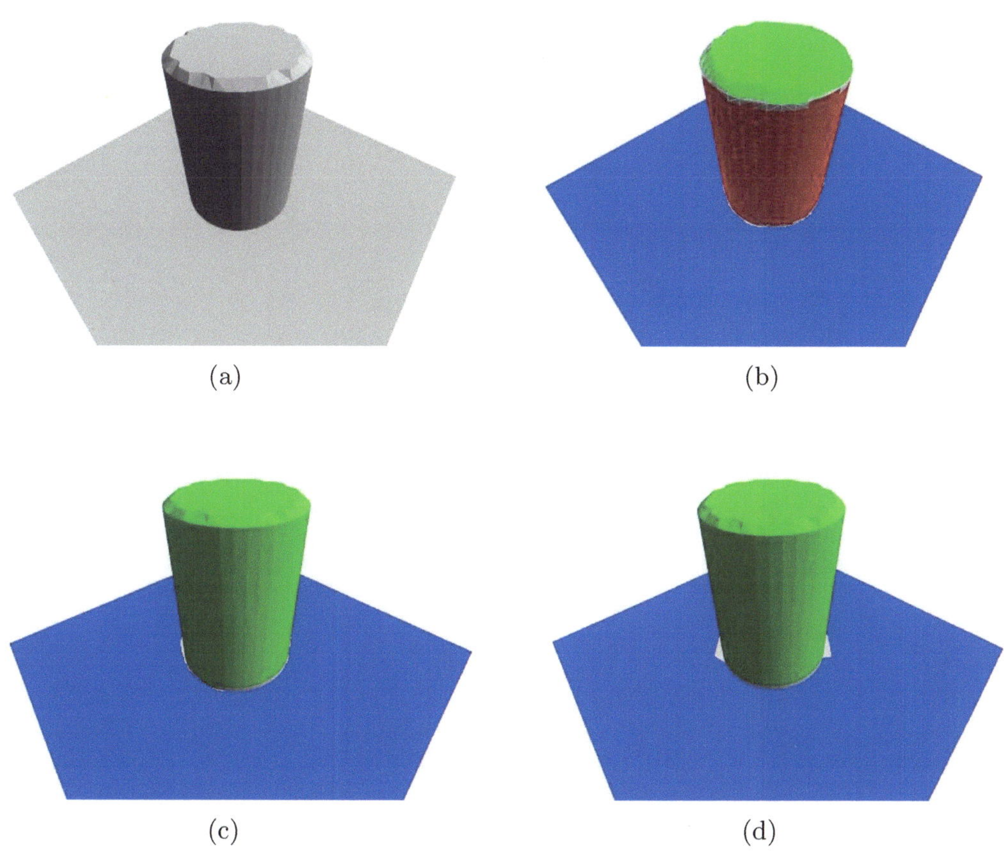

(a)

(b)

(c)

(d)

Figure 7.38: A comparison of segmentation results for a simple scene consisting of cylindrical barrel resting on a floor. (a) Original scene. (b) Segmentation results for Mangan and Whitaker's curvature measure (Mangan and Whitaker, 1999) with three regions: the floor, the barrel side, and the barrel top. (c) Segmentation results for Pulla's curvature measure (Pulla, 2001) with two regions: the floor and the barrel. (d) Segmentation results for our proposed algorithm with two regions: the floor and the barrel.

The second algorithm of interest is in (Pulla, 2001). This algorithm leads to similar results as Mangan and Whitaker, but he employs a different measure of curvature that estimates Gaussian curvature. He uses this measure to also segment along contours of high curvature where the measure does distinguish between positive and negative sense, in contrast to Mangan and Whitaker. So, we have modified Pulla's method slightly to segment along contours of highly negative curvature. This approach is similar to the minima rule definition. We show the results in Fig. 7.38(c), which does agree with the minima rule. This figure shows the segmentation of the scene in to two parts: the barrel and the floor.

Finally, we show the results in Fig. 7.38(d) for our proposed algorithm. This result is similar to the one with Pulla's curvature estimation where the barrel and the floor are the two parts. The question that arises is what is the difference. We find the answer in Fig. 7.39. This scene consists of a floor and a box sitting on the floor. Again, Mangan and Whitaker's algorithm leads to a surface segmentation in Fig. 7.39(b), consisting of six regions where we have one region for each side of the box and the floor. However, with Pulla's curvature method, we see a somewhat unexpected result in Fig. 7.38(c) where we have no segmentation. The algorithm failed to segment the scene. The reason is that the creases at the base of the box that touch the floor form a contour with zero Gaussian curvature. The segmentation algorithm is unable to distinguish these crease from planar regions and as a result does not segment the box from the floor. Our proposed algorithm using principal curvatures overcomes this problem as seen in Fig. 7.39(d). This figure shows a segmentation consisting of the box and the floor. The principal curvatures allow the Fast Marching Watersheds algorithm to differentiate zero Gaussian curvature.

As (Hoffman and Richards, 1984) argue, our algorithm reflects the more likely segmentation that a human observer would choose for the above scenes. We highlight this comparison with a practical example of a disc brake model in Fig. 7.40. This model is a reconstruction from the IVP Ranger Scanner (Sun et al., 2002b). A photograph of the disc brake appears in Fig. 7.40(a), and the corresponding reconstruction appears in Fig. 7.40(b). The output of Mangan and Whitaker is a surface segmentation with four regions in Fig. 7.40(c). Our proposed method, however, yields only two regions in Fig. 7.40(d).

These simple examples demonstrate the minima rule and how that rule differs from previous mesh segmentations, but a potential drawback does exist. We now study this drawback and its implication to the Fast Marching Watersheds algorithm in Figs. 7.41 and 7.42. The L-shaped hammer objects may seem similar at first glance, but a closer inspection reveals that the first object in Fig. 7.41 has a slight indentation at the base of the hammer shaft while the second one in Fig. 7.42 does not. A smooth transition from the handle to the hammer head exists in the second object. The difference in these transitions leads to an ambiguity in the minima rule known as the *part cut* problem (Singh et al., 1999). With 2D silhouettes, we illustrate this ambiguity in Fig. 7.43. The consequence of this ambiguity in our watershed implementation is the results in Figs. 7.41(c) and 7.42(c). The clear minima rule boundary for the first object leads to

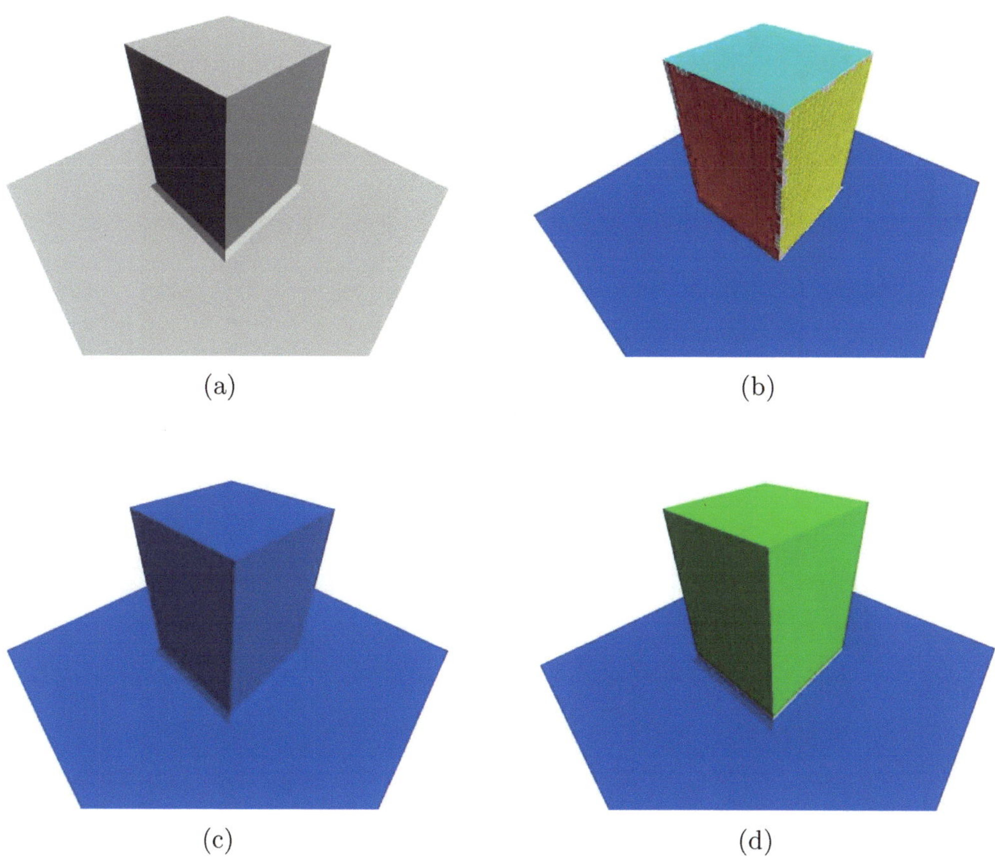

Figure 7.39: A comparison of segmentation results for a simple scene consisting of a box resting on a floor. (a) Original scene. (b) Segmentation results for Mangan and Whitaker's curvature measure (Mangan and Whitaker, 1999) with six regions: the floor, each box side, and the box top. (c) Segmentation results for Pulla's curvature measure (Pulla, 2001) where the algorithm fails to segment the scene. (d) Segmentation results for our proposed algorithm with two regions: the floor and the box.

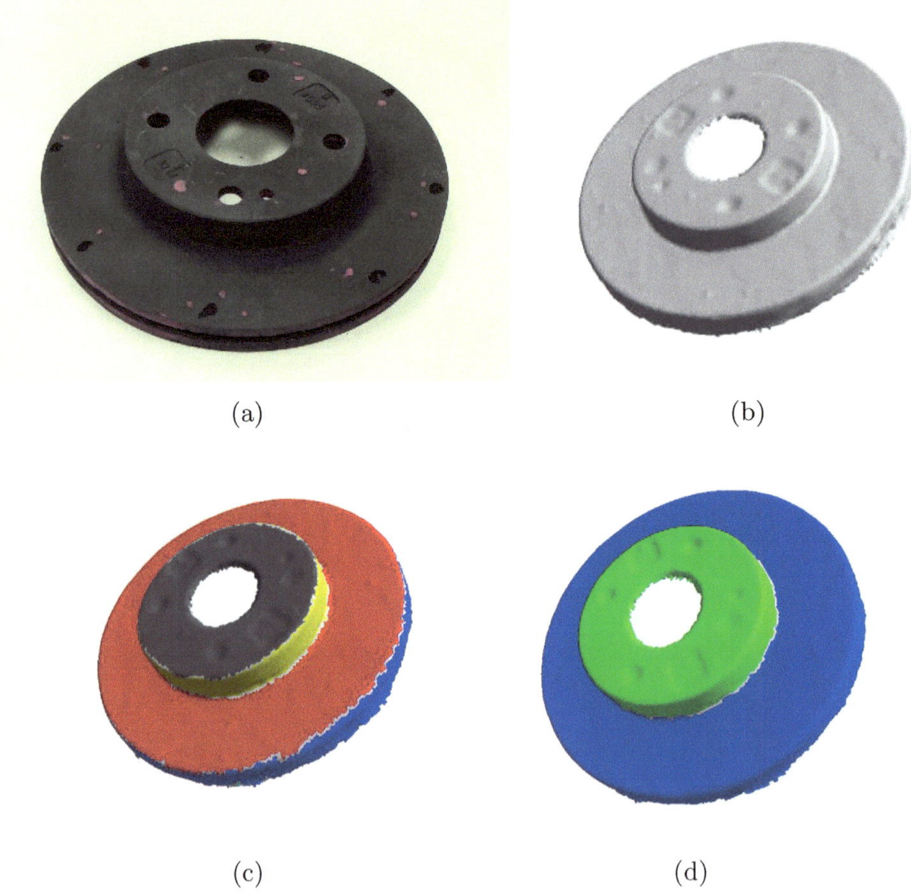

(a)   (b)

(c)   (d)

Figure 7.40: A comparison of segmentation results for a simple scene consisting of a box resting on a floor. (a) Photograph of actual disc brake. (b) Rendering of reconstructed triangle mesh. (c) Segmentation results for Mangan and Whitaker's algorithm with four regions. (d) Segmentation results for the proposed algorithm with two regions—the inner and outer disc.

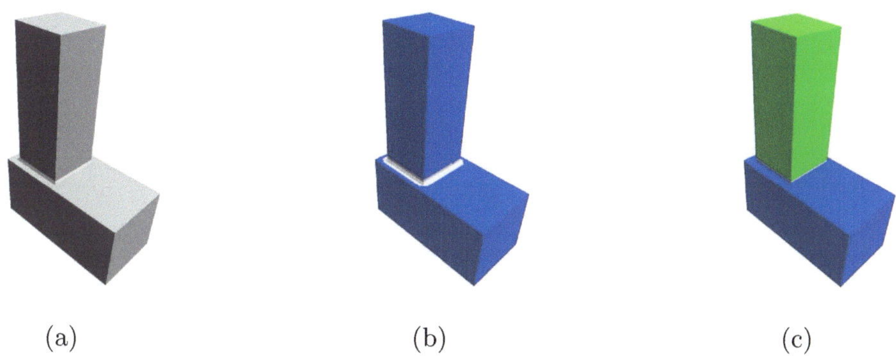

(a)   (b)   (c)

Figure 7.41: The proposed algorithm segments the hammer into two parts: the handle and the head. (a) Rendering of hammer example. (b) Initial threshold of positive curvature values. (c) Results of minima rule segmentation.

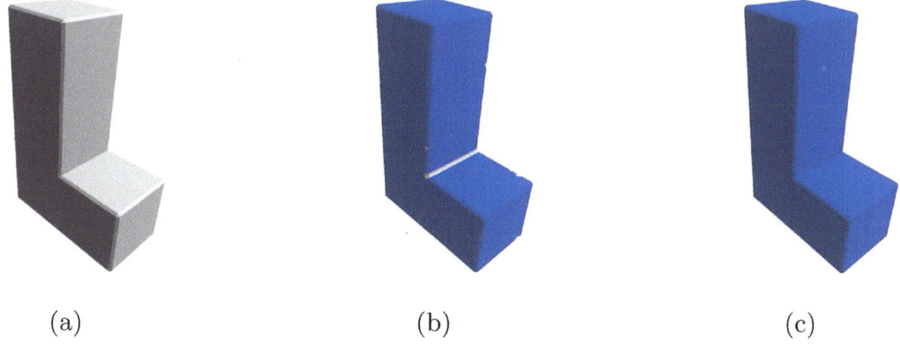

(a)   (b)   (c)

Figure 7.42: The proposed algorithm does not segment the L shape. The segmentation yields only a single object. (a) Rendering of L shape. (b) Initial threshold of positive curvature values. (c) Results of minima rule segmentation.

a segmentation into two parts while the second object results in no segmentation. In the first object, the recess of the vertical block relative to the horizontal one forms a closed negative curvature boundary. For the second one, no clear minima rule boundary is present. As a result, our algorithm leaves the object unsegmented.

We see the part cut problem in real data sets such as in Fig. 7.44. This figure is a triangle mesh from the Polhemus Corporation (www.polhemus.com) and is a reconstruction of a typical chair. In Fig. 7.44(b), the zoom view shows the segmentation of the chair leg and the support bar. These parts follow the minima rule. By contrast, Figs. 7.44(c) and 7.44(d), show the ambiguity of the part cut problem. For Fig. 7.44(c), the algorithm does not segment the chair seat and the connection to the leg into two separate parts. Likewise, in Fig. 7.44(d), the algorithm does not cut the chair back into a side and top segment. This drawback is not necessarily a problem with the

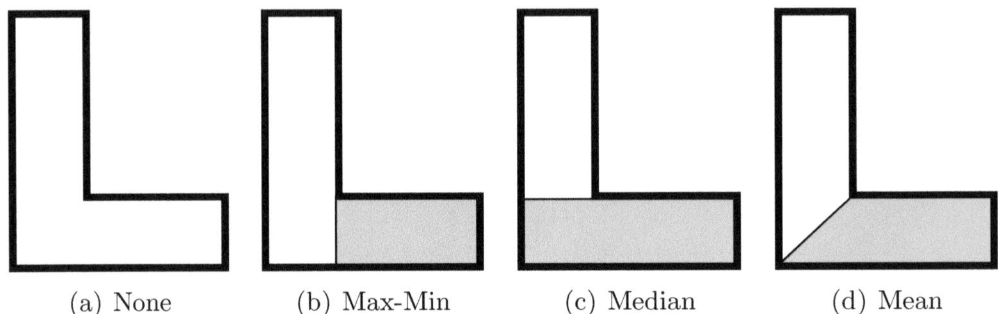

(a) None (b) Max-Min (c) Median (d) Mean

Figure 7.43: Possible part cuts for L shape.

Fast Marching Watersheds algorithm, but it is an ambiguity in the minima rule that is important to understand.

In this section, we have compared our proposed algorithm to that of Mangan and Whitaker (Mangan and Whitaker, 1999) and Pulla (Pulla, 2001). Also, we have demonstrated the part cut ambiguity that arises with the minima rule and how that effects our segmentation method. This investigation serves as a qualitative analysis of the Fast Marching Watersheds algorithm. In the next section, we seek a more quantitative investigation.

7.3.2 Quantitative Analysis

The previous examples show the capabilities of the Fast Marching Watersheds algorithm but do not provide a qualitative analysis of the algorithm parameters. Figs. 7.45, 7.46, and 7.47 reflect a more focused investigation to study the choices for α and k user parameters.

The first parameter that we consider is the threshold α in Eq. (5.18). Recall that a threshold of $t > 0$ is the ideal case for the minima rule, but we introduced α because we must offset the threshold to a slightly negative value to handle practical conditions. If we choose the mug, the disc brake, and the chair from the previous sections and analyze their curvatures, we see why α is necessary. In Fig. 7.45(a), we plot the negative principal curvatures for each of these three examples. These graphs are bin plots where the x-axis denotes the negative curvature bin while the y-axis denotes the percentage of vertices with that particular curvature value. These plots are indicative of the general trend that we see in most triangle meshes that are reconstructions of objects and scenes. A significant number of vertices exhibit curvature very close to zero with a rapid decay as one moves away from zero. Most likely, the values near zero should actually be zero, but errors in estimation lead to slightly negative values. Therefore, α allows the user to group these slightly negative values into the zero curvature bin.

The question that we face is what value of α provides the most robust results. If we again use the mug, we gain insight to this question with the plots in Fig. 7.46. These two graphs study the effects of varying α through the range $[0, 1]$. When $\alpha = 0$, we use

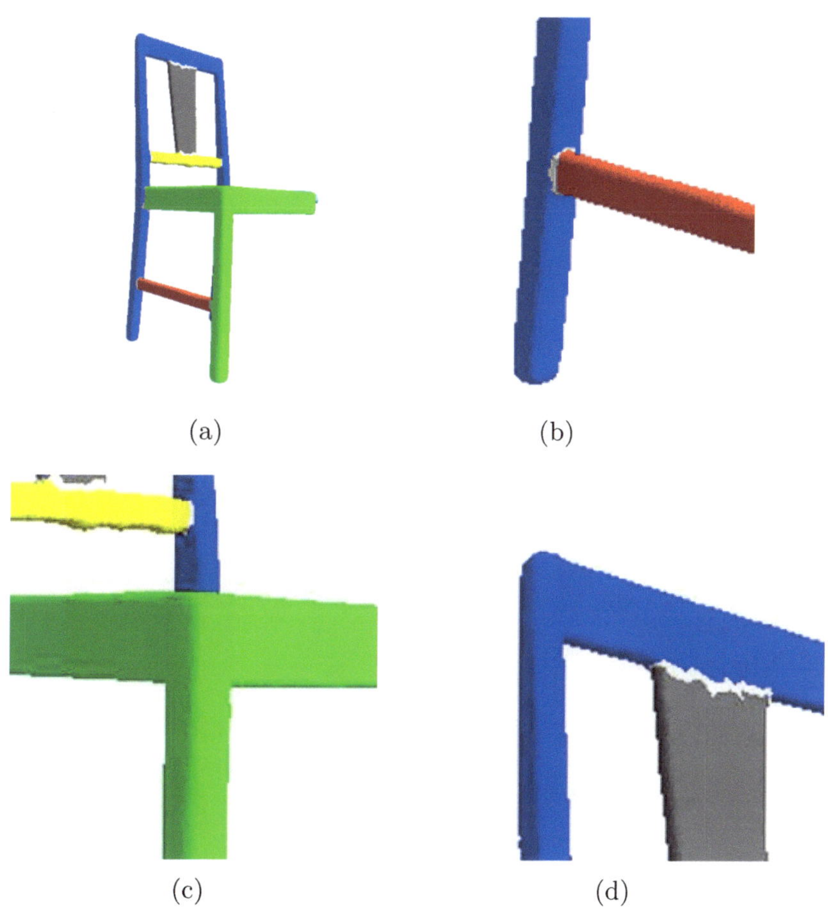

Figure 7.44: Fast Marching Watersheds segmentation of the Polhemus chair that illustrates the potential drawback of the minima rule. (a) Segmentation of the chair. (b) Zoom view of leg and support bar showing minima rule segmentation. (c) Zoom view of seat and leg intersection. The algorithm does not segment at this intersection as a result of the part cut ambiguity. (d) Zoom view of the top and side support for the back of the chair. As with (c), the algorithm does not segment at this intersection.

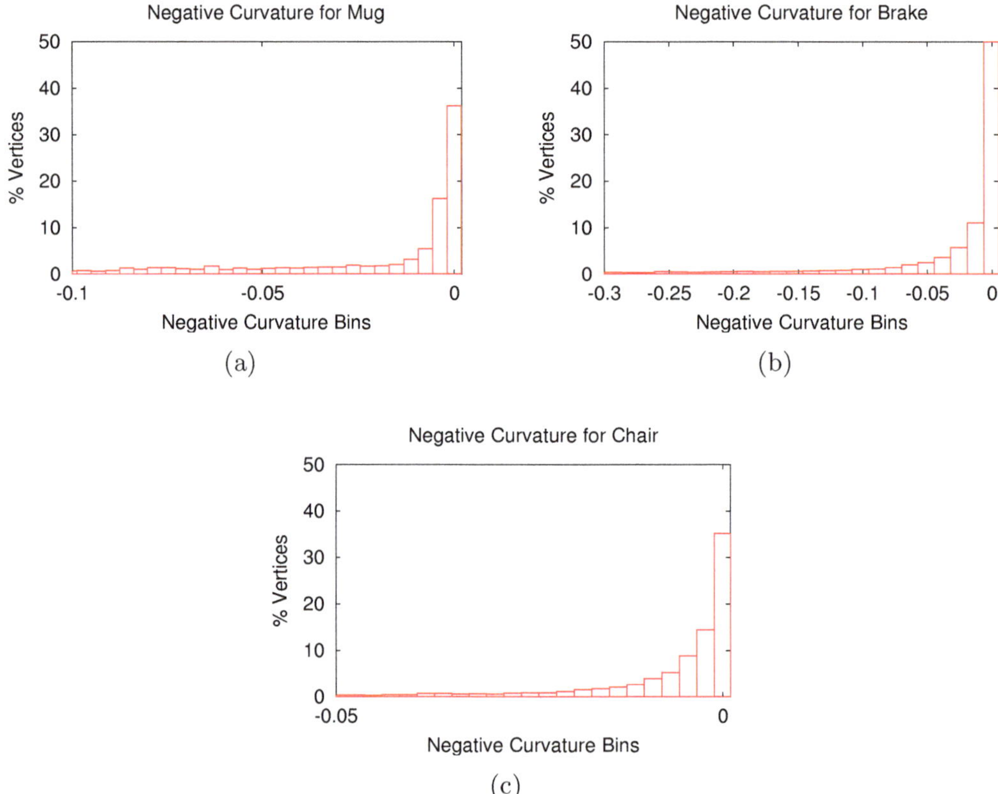

Figure 7.45: Distribution of negative principal curvatures for (a) the mug, (b) the Ranger disc brake, and (c) the Polhemus chair.

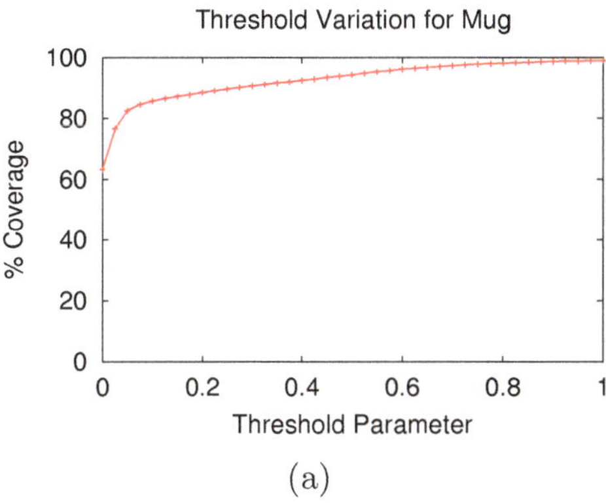

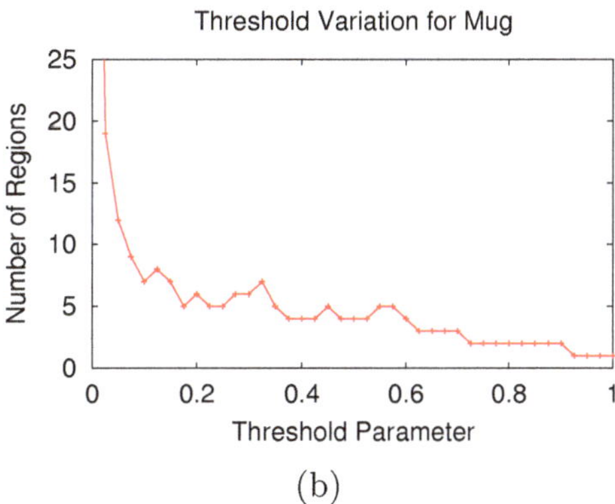

(a) (b)

Figure 7.46: These plots study the variation of α to understand the threshold operation. The mesh under consideration is the coffee mug. (a) The percentage of the mug that is above the threshold value as a function of α. (b) The number of unique connected regions as a function of α.

the ideal threshold of $t > 0$. When $\alpha = 1$, we use the average negative curvature value as the threshold. The first graph in Fig. 7.46(a) shows the percentage of the mesh that we segment as α varies. From this graph, we see that 60% of the mug is segmented for $\alpha = 0$ while almost 100% of the mug is segmented for $\alpha = 1$. This result suggests that our definition of α is an appropriate parameter to control the threshold.

The second graph in Fig. 7.46(b) shows the variation of the number of connected regions as a function of α. As with the first graph, the range of α yields diverse results. When $\alpha = 0$, we generate over 25 different unique regions for potential mug parts. Obviously, this number is way too many based on our perceptual experience. When $\alpha = 1$, the number of connected regions is one as we might expect from the data in the first graph. With almost 100% of the mug segmented a natural conclusion is that the segmented regions are probably connected. The interesting feature of both of these graphs is the "knee" in the plots that occur around $\alpha = 0.1$. These transitions are typical for the data sets we have tested and usually occur between $\alpha = 0.05$ and $\alpha = 0.2$. Thus, with the evidence from these plots and our experience with the data sets, we recommend $\alpha = 0.3$ as the most reliable choice for stable results. With $\alpha = 0.3$, we include the capture the knee transition where meaningful threshold results occur. However, since we further process the threshold using morphological operations, the algorithm is not overly sensitive to this choice, and other choices are useful depending on the needs of an application.

Although α is the primary control of the Fast Marching Watersheds algorithm, the morphological operations also allow the user to control the algorithm, as well. The parameter for these operations is k, which is the size of the D_k structuring element. The sampling density of the mesh tightly couples the choice for k to that for α. As a result, if we fix one value, we most likely have fixed the other value, too. For the sequence in Fig. 7.47, we choose $\alpha = 0.3$ per our previous discussion, and we vary the value of k.

The initial threshold without morphological processing appears in Fig. 7.47(a). This subfigure shows nine different connected regions, which is more than we expect. Particulary, this threshold leaves small unconnected islands around the base of the handle. To improve this result, we first apply the opening and closing operations with $k = 1$ in Fig. 7.47(b). Morphology processing either eliminates the islands or joins them to a nearby region. This choice for k in conjunction with our choice for α leads to an appropriate marker set. As shown in the subfigure, the marker set now consists of five unique regions. Four of the regions correspond to the perceptual parts that we seek in a segmentation such as the handle, the cup, the base cusp, and the bottom. However, we still have a small island region at the base of the handle. To eliminate this island, we can not continue to increase k. Fig. 7.47(c) demonstrates the effect of applying a structuring element with $k = 2$. This subfigure shows only two unique regions. We have lost the markers for our desired perceptual features. The situation continues to grow worse when we further increase $k = 3$ in Fig. 7.47(d).

What we conclude is that $k = 1$ provides the most useful marker set that minimizes the amount of over segmentation. Without morphological operations, the mug segmentation would have nine marker regions, which is five more than the desired four perceptual regions. Our post processing with the Part Saliency Metric would have to account for these five additional regions. With the morphological processing, on the other hand, we have one region at the base of the handle that is an over segmentation. Thus, we have improved our marker set and reduced the burden on the Part Saliency algorithm. Our experience with other data sets besides the mug leads to similar results that $k = 1$ works best when we choose $\alpha = 0.3$. The $k = 1$ structuring element essentially bridges gaps and removes isthmuses that one to two vertices wide. If we have a fairly good curvature estimation and a fairly good threshold, data irregularities of one to two vertices are expected.

7.3.3   Timing Performance

Finally, the graph in Fig. 7.48 shows the timing performance of the proposed algorithm for the mug data set. To demonstrate the performance, we vary the resolution of the mug mesh with a mesh simplification algorithm. The initial mug consists of $10,555$ vertices and $20,910$ triangles. We reduce this size in increments of 10% on the x-axis. The y-axis is the processing time required. We plot four curves on the graph. Three of the curves (*Curvature*, *Threshold*, and *Watershed*) represent the computational time for each individual module of the algorithm. The *Threshold* time includes the morphology operations, as well. The *Total* time is the summation of the three modules. The curves appear close to linear except for the jump in the Threshold curve from 50% to 60%. This jump occurs because below 50% the mug does not require morphology operations to clean up the threshold while above 60% the opening and closing operations are necessary. The jump reflects the performance cost for the morphology operations. We also note that the fast marching module of the algorithm contributes very little to the overall timing while the curvature module contributes the most. Although the

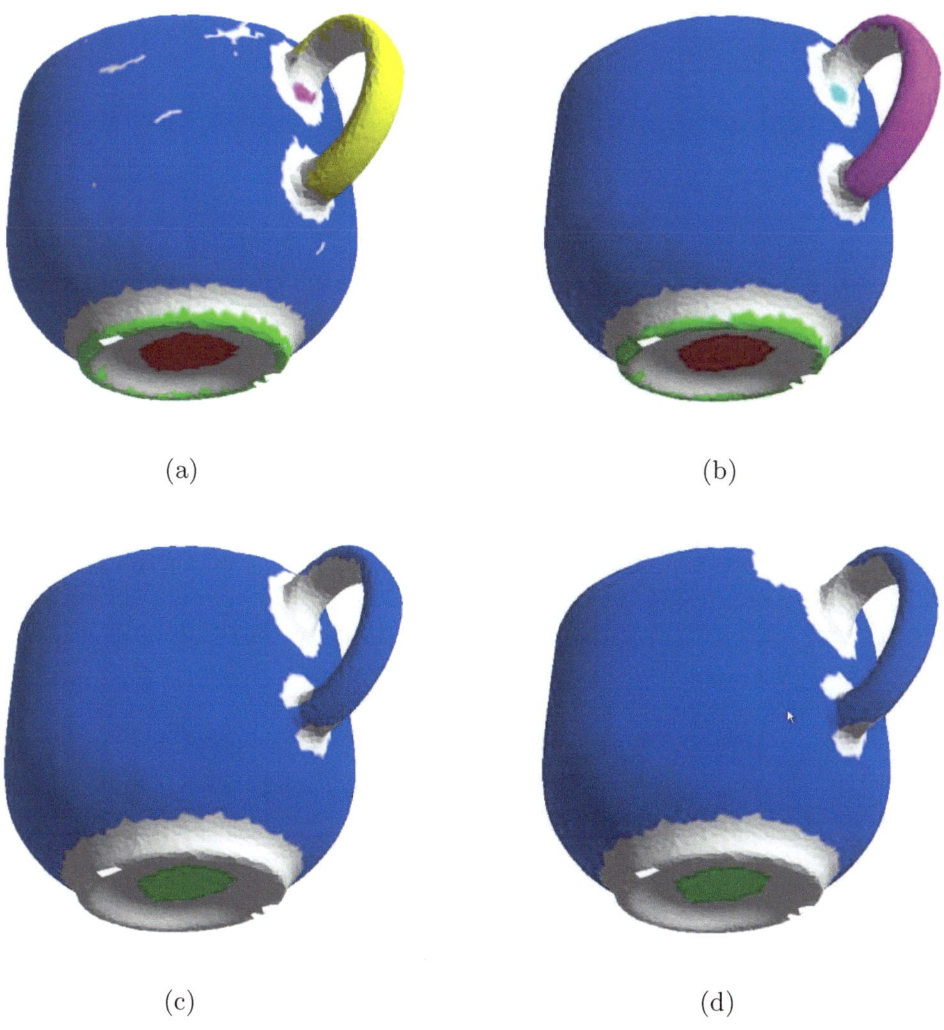

Figure 7.47: For these examples, we vary the size of the structuring element D_k for the morphology operations. (a) Initial threshold of the mug with $\alpha = 0.3$ results in 25 unique marker regions. (b) Set $k = 1$ and apply opening and closing operations results in five unique marker regions. (c) Set $k = 2$ leads to two marker regions. (d) Set $k = 3$ also leads to just two marker regions.

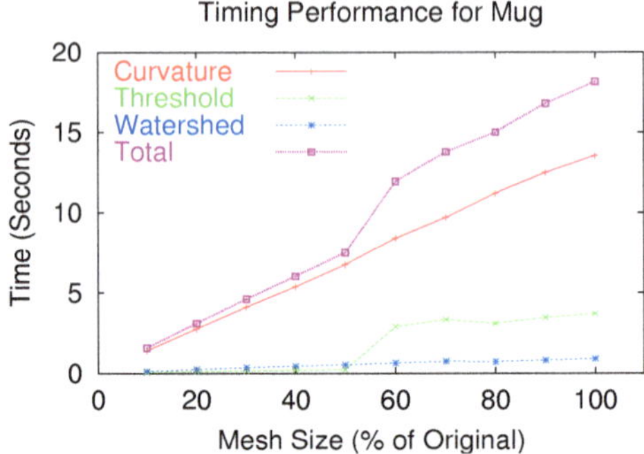

Figure 7.48: Timing performance plot of the code modules for the Fast Marching Watersheds as a function of mesh size.

$O(n \log n)$ nature of fast marching is inherently fast, the fact that thresholding alone segments typically 70 to 90% of the mesh improves the performance, as well.

7.4 Part Saliency Metric

With the results from the previous two sections, we are able to compute the principal curvatures at each vertex of a triangle mesh and then segment the mesh along lines of negative curvature minima. Our implementation of the minima rule decomposes a mesh into visual parts. To evaluate the quality of the visual parts, we have developed the Part Saliency Metric algorithm. In this section, we analyze this algorithm. We assume that we have a mesh that has been segmented into visual parts according to the minima rule. These parts however may or may not be visually salient to a human observer. The algorithm measures the perceptual importance of each part based on a theory of human vision. We first evaluate in a qualitative manner how the algorithm computes saliency, and then we compare this measure to other possible measures. Finally we conclude with an evaluation of the timing performance of the algorithm.

7.4.1 Qualitative Analysis

Recall the sequence of part merges for the mug in Figs. 6.1 and 6.2. For this sequence, the Part Saliency Metric controls the merge of the least salient parts as we move from a segmentation of five parts to a single part. We claimed in these previous figures that the sequence followed the measured saliency values, but we did not present those values. We now do so in Figs. 7.49 and 7.50.

The bar charts in these figures demonstrate the relative salience of each part. In the charts, we have ordered the bars such that the most salient parts are on the left of the graph while the least are on the right. The legends of each graph show the corresponding part of the mug. The cup, handle, and bottom parts are fairly obvious while the knob and base may not be. The latter two parts are an oversegmentation.

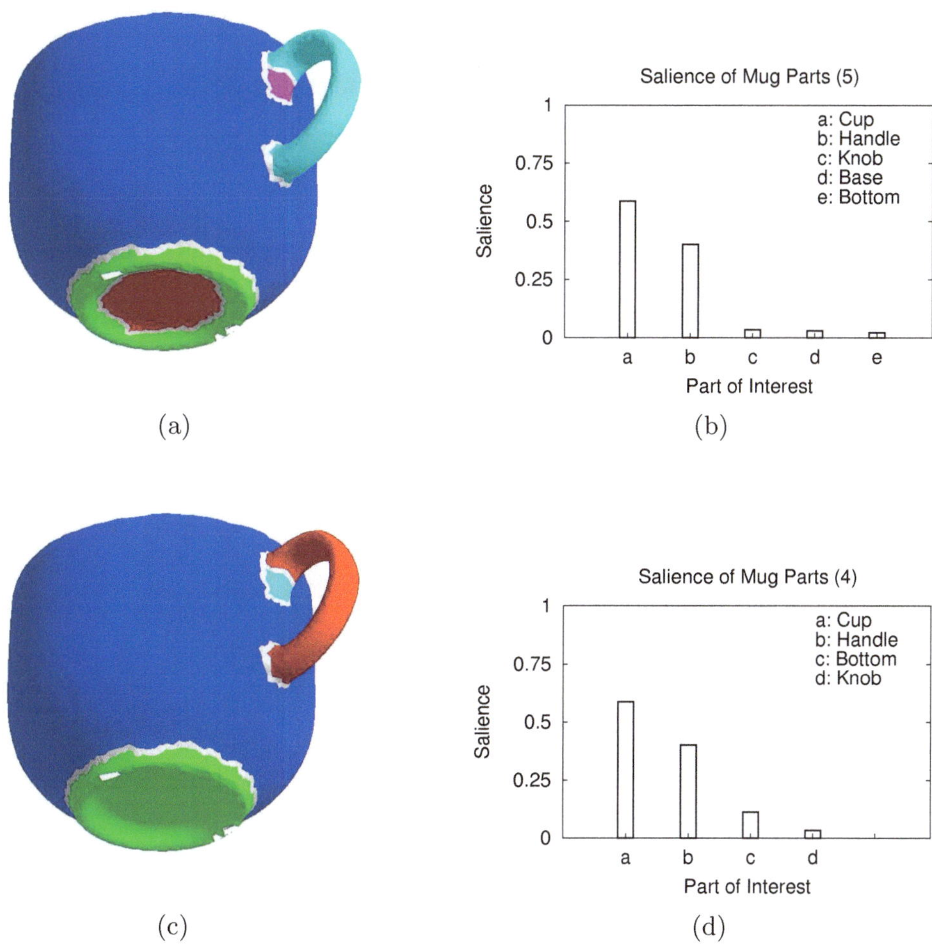

(a)   (b)

(c)   (d)

Figure 7.49: These bar graphs show the saliency metric for each part of the mug. The progression sequence shows the metric changes as we simplify the number of parts for the mug. (a) Five parts: cup, handle, base, bottom, and knob. (b) Saliency metrics for these five parts. (c) Four parts: bottom merges with base. (d) Saliency metrics for these four parts.

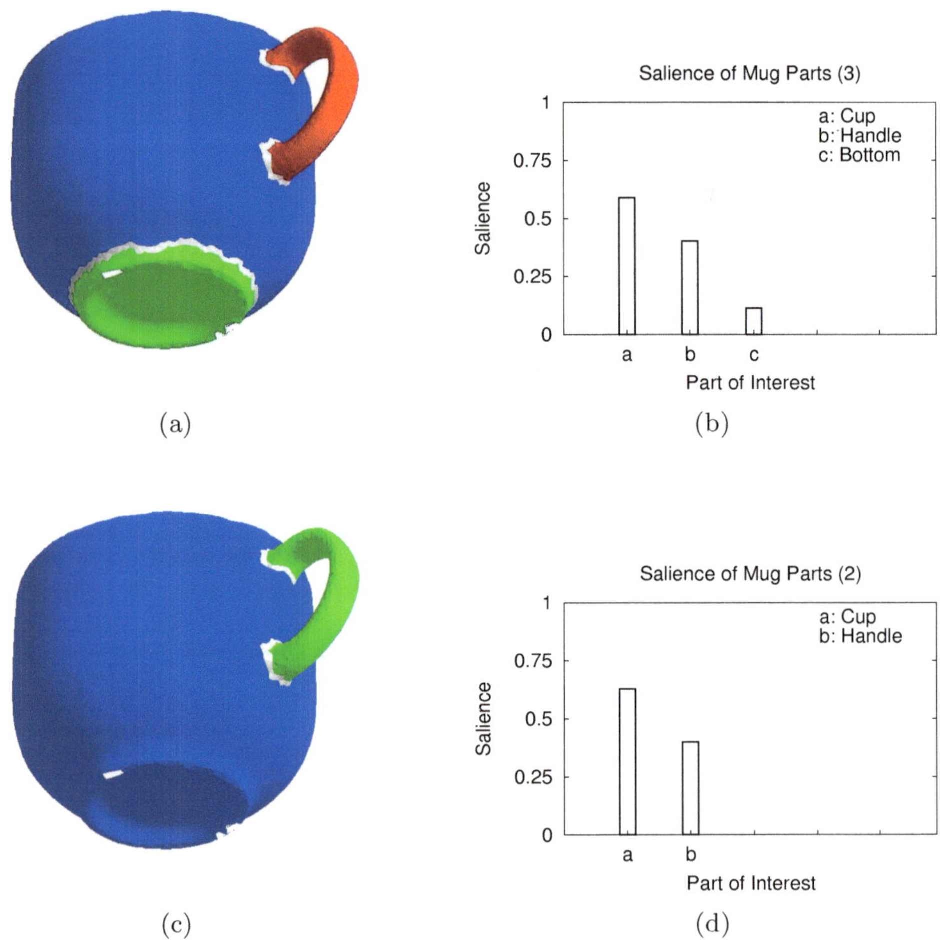

(a)   (b)

(c)   (d)

Figure 7.50: These bar graphs show the saliency metric for each part of the mug. The progression sequence shows the metric changes as we simplify the number of parts for the mug. (a) Three parts: cup, handle, and base. (b) Saliency metrics for these three parts. (c) Two parts: bottom merges with cup. (d) Saliency metrics for these two parts.

The knob is the small portion at the top intersection of the handle with the cup. The base is the flat region underneath the mug. From Fig. 7.49(b), we see that the knob, the base, and the bottom are the least salient parts of the initial segmentation. They have almost equivalent values. The algorithm selects the bottom as the least salient and merges it with the base. This operation results in the new bottom, which exhibits a larger salience than either the base or original bottom alone as shown in Fig. 7.49(d). The merger clearly leaves the knob as the least salient part. So, if the merge algorithm continues, the knob joins the cup with the results in Fig. 7.50(b). Finally, we merge the bottom with the cup in Fig. 7.50(d). The measured saliency values drive the merge progression of least salient parts to more salient ones.

As these bar graphs show, the importance of the Part Saliency Metric is to evaluate the quality of each part as a part itself. For example, the knob, the base, and the bottom in the original segmentation are not visually significant parts, and we have a measure that quantifies this notion. Further, if we merge the base and the bottom, we see that these parts together form a more visually significant part, and the measure captures this change, as well. This ability to measure the quality of each part is the contribution of this algorithm.

We continue to explore the saliency measure in Figs. 7.51 through 7.57. Here, the data set is an automotive distributor cap. The triangle mesh for this model consists of 65K vertices with 130K triangles. A rendering of the model appears in Fig. 7.51(a) with the part decomposition in Fig. 7.51(b). This decomposition consists of 20 different parts. The saliency algorithm generates an adjacency graph that shows the interconnection of these parts. This graph appears in Fig. 7.52. The nodes of the graph represent the parts while the edges define the part connections. We have manually assigned the node labels to describe the appropriate part of the distributor cap. This graph benefits other computer vision tasks such as object recognition.

Once we have a graph representation of the segmentation, we can also apply the four-coloring algorithm as in Fig. 7.53. The segmentation in Fig. 7.53(a) is the original segmentation where a simple method alternates among a palette of 16 different colors. For some segmentations, this method is adequate and usually does not lead to ambiguities for full color visualizations such as on a computer monitor. However, in grey scale such as one may find with a laser printer, the 16 colors are visually difficult to distinguish for adjacent parts as in Fig. 7.53(c). With a palette of four colors and the four-coloring algorithm, we are able to improve the color labels as in Fig. 7.53(b). With these labels, the colors do not simply cycle among the palette, but rather the algorithm intelligently selects a color from the palette so that adjacent parts do not share a common color. This approach improves the contrasts for visualizations in grey scale such as Fig. 7.53(d). The four-color labeling of the segmentations is often more visually pleasing for output in grey scale.

The previous results for the distributor cap demonstrate the adjacency graph and the four-coloring problem for the Part Saliency Metric, but the heart of the algorithm is the saliency measure as shown previously with the mug data set. The bar graph in Fig. 7.54 shows the saliency values for the distributor cap parts. Note that this

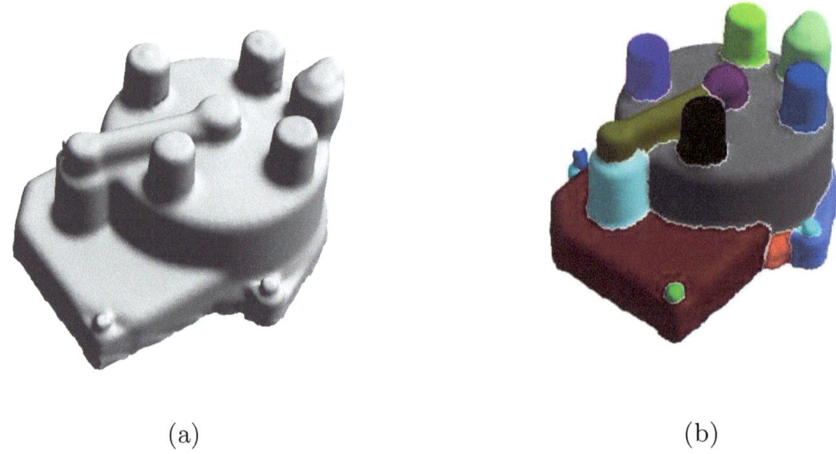

(a)   (b)

Figure 7.51: These screen shots show the segmentation of the distributor cap model, which consists of 65, 397 vertices and 129, 849 triangles. (a) Rendered view showing the shape of the model. (b) Colored labeling of the segmentation.

graph has values for 22 parts whereas the results in Figs. 7.51 and 7.52 are for only 20 parts. The two extra parts are **u** and **v**. As the bars indicate, these parts have the smallest values, and thus the algorithm merges them with other parts to achieve the final segmentation in Fig. 7.51(b). We qualitatively call parts **u** and **v** "bad" parts. The letter designation for each part and its corresponding description are in Table 7.6.

Fig. 7.55 shows the physical location of parts **u** and **v** on the distributor cap. They occur at the rear where some irregularities in the surface of the model occur. These irregularities are most likely measurement errors that lead to problems in surface reconstruction. As the zoom views in these figures show, neither part bounds a significant volume, nor do they protrude much from the model. These factors lead to small saliency values. By comparison, the other parts of the distributor cap have significantly larger

Table 7.5: Part letter designations for distributor cap model with corresponding descriptions.

a	Cap	i	Base D	p	Screw A
b	Base	**j**	Base F	**q**	Link Base
c	Base A	**k**	Plug B	**r**	Base B
d	Link Tube	**l**	Plug D	**s**	Screw C
e	Plug E	**m**	Screw B	**t**	Base C
f	Plug C	**n**	Base E	**u**	Bad Part A
g	Link Cap	**o**	Screw D	**v**	Bad Part B
h	Plug A				

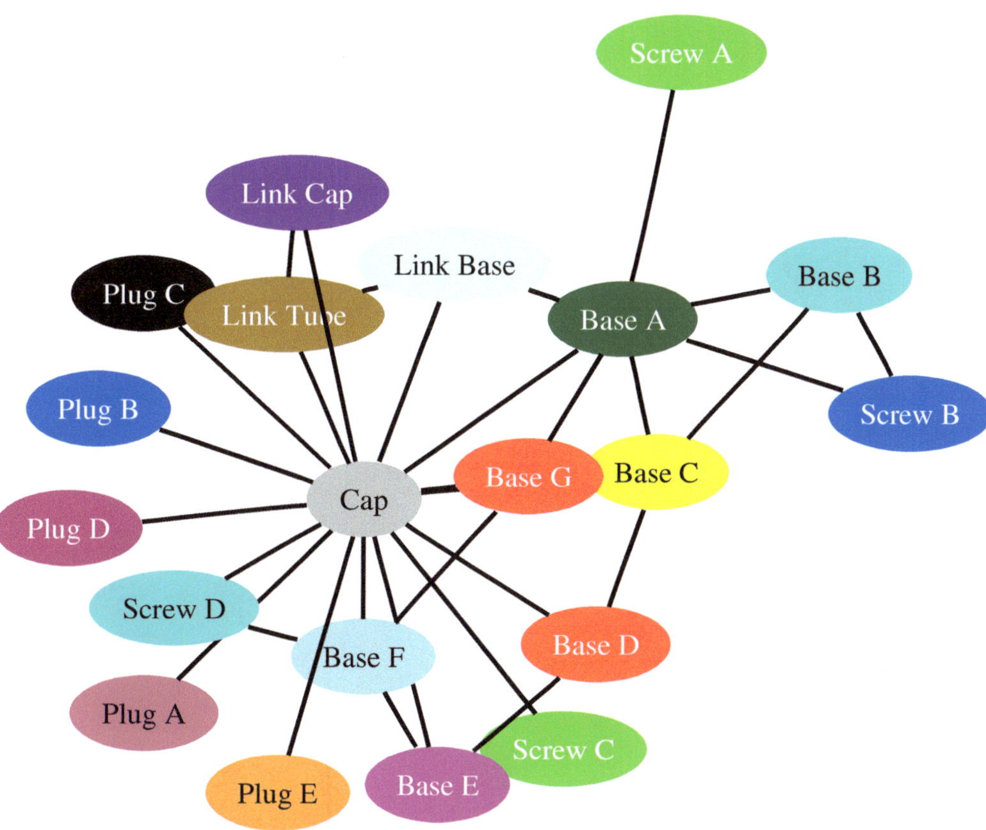

Figure 7.52: The above undirected graph is the part adjacency graph for the segmentation of the distributor cap model. The nodes of the graph represent the parts of the model while the edges represent the connections between parts. We manually assign the node labels.

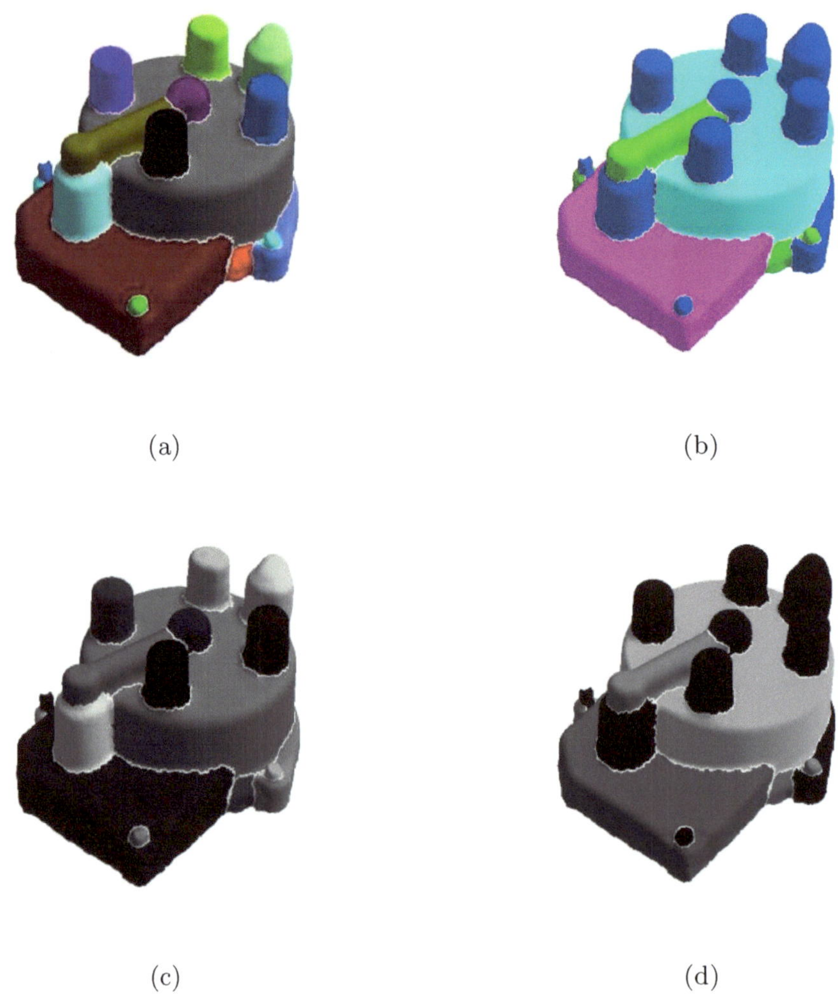

(a)   (b)

(c)   (d)

Figure 7.53: This coloring of the distributor cap segmentation is the result of a four-color graph algorithm. Such an algorithm improves the distinction of the region colors, especially for black and white contrast. (a) Color labels that alternate among a palette of 16 different colors. (b) Four-color labels where adjacent regions never share a common color. (c) Grey scale version of (a). (d) Grey scale version of (b).

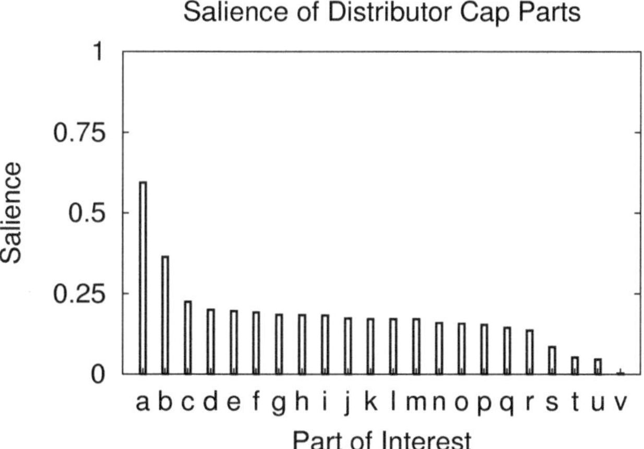

Figure 7.54: The bar graph shows the relative saliency of each part from the initial segmentation of the distributor cap. This segmentation is before merging of the least salient parts. The parts with the smallest saliency are parts **u** and **v**, which correspond to the "bad" parts.

values. Consider the largest part **a**, which is the center cap from where the other parts extend. This cap occupies the most volume and is thus the most significant part of the model. The advantage of the proposed measure, however, is when we consider parts such as the screw **s**. A zoom view of this screw appears in Fig. 7.56. In terms of surface area, the screw seems quite small in comparison to **u** and **v**. Indeed it is small, but in terms of salience as shown in the bar graph, the screw is much larger. This result agrees with our visual perception since most human observers would select the screw over parts **u** and **v** as salient features of the distributor cap.

The third least salient part in the graph of Fig. 7.54 is part **t**. This part is a small protrusion at the base of the distributor cap. The zoom view in Fig. 7.57 shows a better view of this part. From the bar graph, we know that this part is only slightly more salient than **u** and **v**, but less salient than the screw **s**. The question is whether or not this part is a "good" or a "bad" part. This decision is essentially a threshold of the saliency measure. Since this threshold differs for different human observers, establishing a unique criteria is not trivial—and may not be possible. Thus, we pose this question as an area of future research. The control of the merge algorithm requires the specification of a minimum saliency value to define "good" parts.

7.4.2   Quantitative Comparison

We now investigate the Part Saliency Metric in a more quantitative manner. In particular, we compare the results of the saliency measure to two other measures for evaluating the quality of the segmentation. The results appear in Figs. 7.58 through 7.62. The first measure that we compare is the surface area of the part. To compute this measure, we simply sum the areas for each triangle associated with a particular part. Sun et al. (Sun

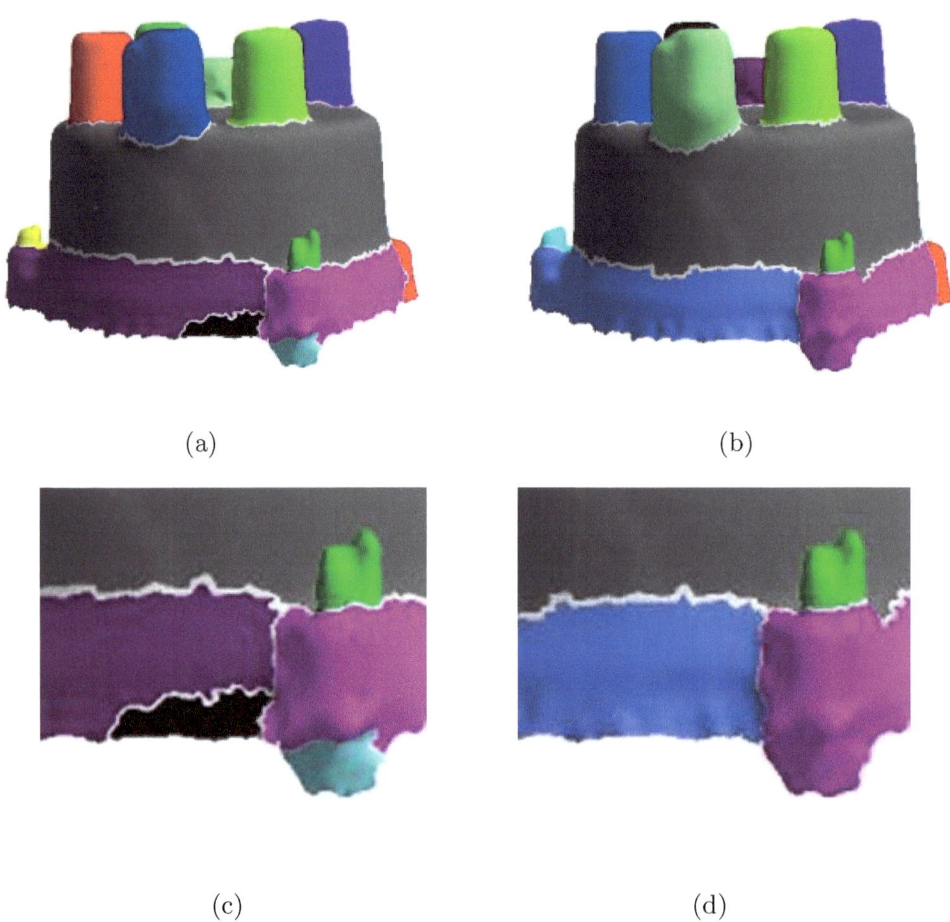

(a) (b)

(c) (d)

Figure 7.55: The views on the left show "bad" parts from the segmentation of the distributor cap. The views on the right show how evaluation with the saliency metric allows us to merge these parts to improve the segmentation. (a) Rear view of distributor cap after initial segmentation with 22 parts. (b) Rear view of distributor cap after merging least salient parts with 20 parts. (c) Zoom view of region of interest in (a) showing two "bad" parts: **u** and **v**. (d) Zoom view of corresponding region in (b).

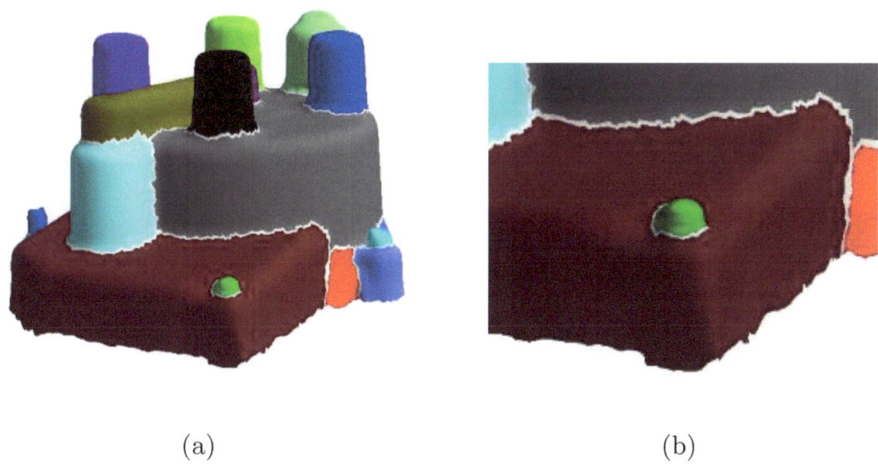

(a)   (b)

Figure 7.56: The salience metric allows the algorithm to preserve salient, or "good", such as this screw, which is part **s**. (a) The broad view that shows the context of the zoom view in (b). (b) A zoom view of the screw that protrudes from the base of the model.

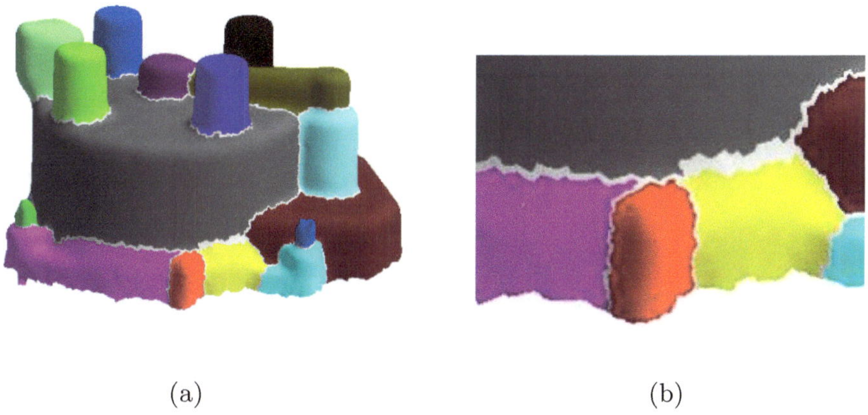

(a)   (b)

Figure 7.57: The base of the distributor cap consists of a few different regions. One of those regions is a small protrusion, which is part **t**. (c) This view shows the context for the view in (d). (d) A zoom view of the base that protrudes from the model.

et al., 2002a) use this approach to improve their segmentations. The approach assumes that regions that have a small area relative to some threshold are oversegmentations. The second measure is the watershed depth of the part. To compute this measure, we find the minima watershed height for a part and difference this value with the maximum height along the boundary of the part. With our curvature definition for watershed height, this approach means that we find the maximum principal curvature for a part and then the minimum principal curvature along a part boundary. The difference between these values is the depth of the part. The assumption is that parts with small depths are oversegmentations. This method is common in the literature (Mangan and Whitaker, 1999; Rettmann et al., 2002). We have found these methods to be *ad hoc* and are not as robust as the visual saliency measure that we have proposed. The results that follow support this argument.

To begin our comparison, we first create a set of synthetic triangle meshes to highlight the variations between the different methods. These objects appear in Fig. 7.58 and their corresponding segmentations appear in Fig. 7.59. These objects consist of two parts: a lower box and an upper box. The physical dimensions of the lower box are the same among the four objects while dimensions of the upper boxes vary. The variation is such that the visual saliency of the upper box increases from left to right and top to bottom in the figure. We label the four objects as shown.

We have carefully defined the dimensions of the upper boxes to highlight the differences of the three quality measures. For example, the doormat and tall box object have roughly the same surface area while the pancake and short box have approximately half that surface area. For the watershed depth, the doormat object has almost zero depth while the pancake, short box, and tall box have nearly identical maximum and minimum curvature values. However, by varying the bounded volume, the part protrusion, and the boundary strength, each of the parts have very different visual saliency values.

We begin with a comparison for the part area in Fig. 7.60. These measures result in almost equivalent values for the doormat and the tall box, and their values are the largest of the four objects. The pancake and short box objects, on the other hand, have the smallest values. These values reflect our careful creation of the synthetic data set. With this measure, one might conclude that the doormat is a better part segmentation than either the pancake or short box. This conclusion disagrees with our visual perception of the parts, however. The doormat is almost flush to the top surface of the lower box and has very little visual importance, but the pancake and the short box are readily discernable as they protrude from the lower box. They perceptually stick out more than the doormat. The part area does not necessarily capture our visual perception of the parts. A large region in terms of area does not directly imply a "good" part. The "bad" parts from the distributor cap in Fig. 7.55(c) are a practical example.

If part area does not quite meet our needs, then we next consider the part depth as a quality measure. Since we are using a watershed algorithm to generate our segmentations, we can define a watershed depth to evaluate the quality of the segmentation regions. For our synthetic data sets, the results appear in Fig. 7.61. Interestingly, the doormat receives the lowest value, which was not the case with part area previously.

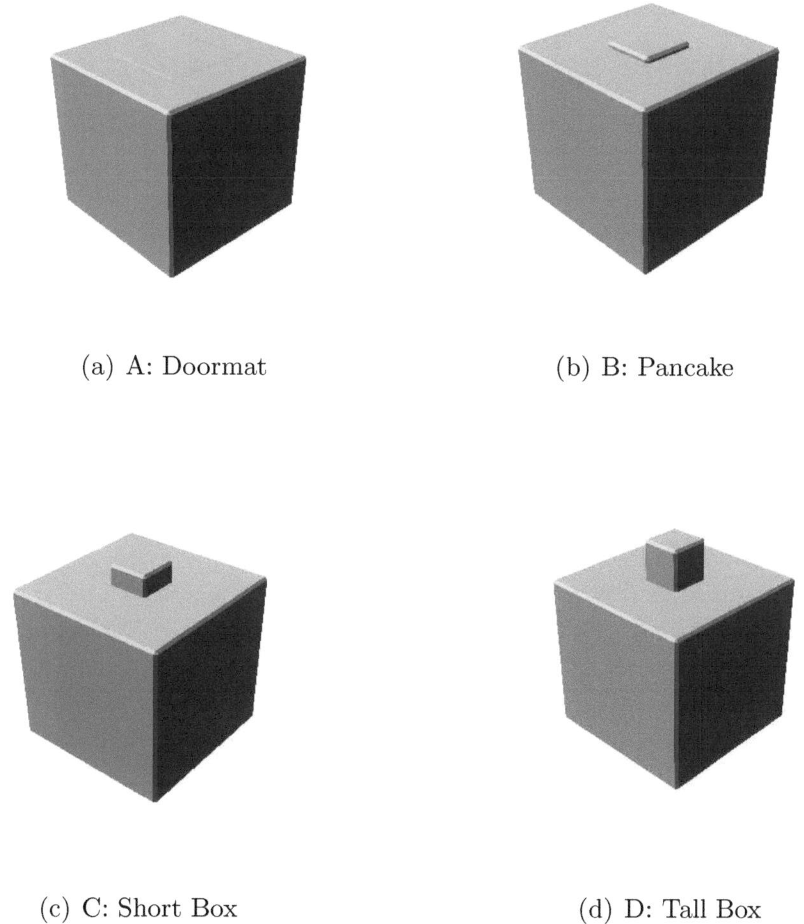

(a) A: Doormat

(b) B: Pancake

(c) C: Short Box

(d) D: Tall Box

Figure 7.58: These objects consist of two parts. The lower box is the same size for each of the objects. The upper box has varying dimensions to simulate different visual saliency values. (a) The doormat object where the top part is almost flush with lower box. (b) The pancake object where the top part protrudes only slightly. (c) The short box object where the top part has significant volume and protrusion. (d) The tall box object where the top part extends like an obelisk from the lower part.

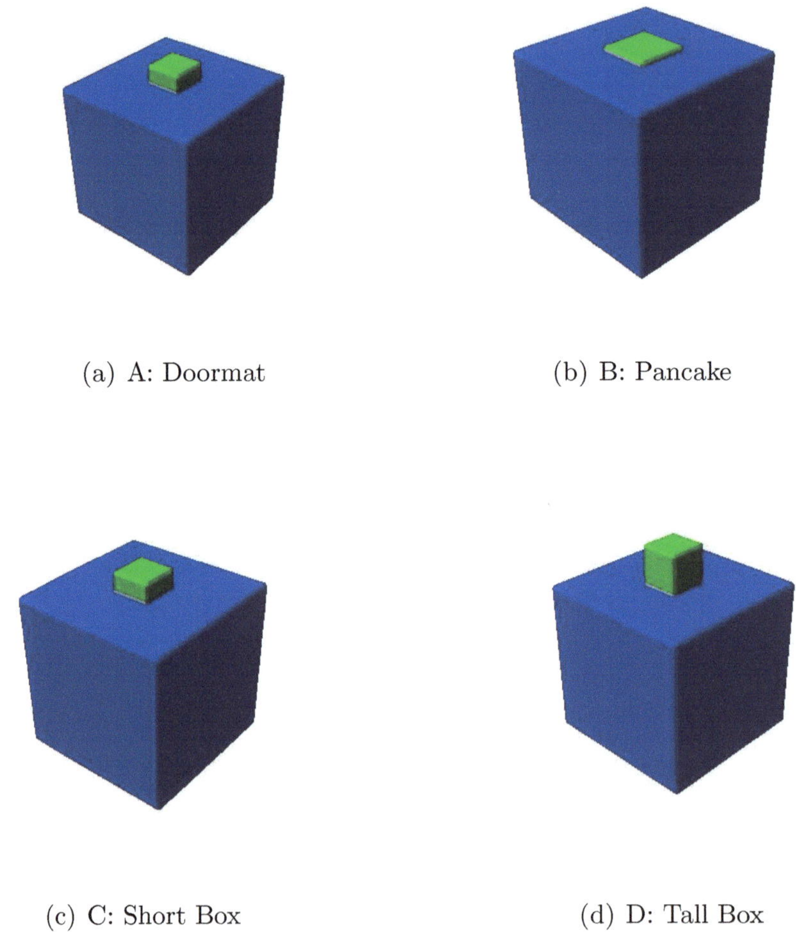

(a) A: Doormat

(b) B: Pancake

(c) C: Short Box

(d) D: Tall Box

Figure 7.59: These segmentations show the parts for the objects of the previous figures. Each object consists of two parts. (a) The doormat object where the top part is almost flush with lower box. (b) The pancake object where the top part protrudes only slightly. (c) The short box object where the top part has significant volume and protrusion. (d) The tall box object where the top part extends like an obelisk from the lower part.

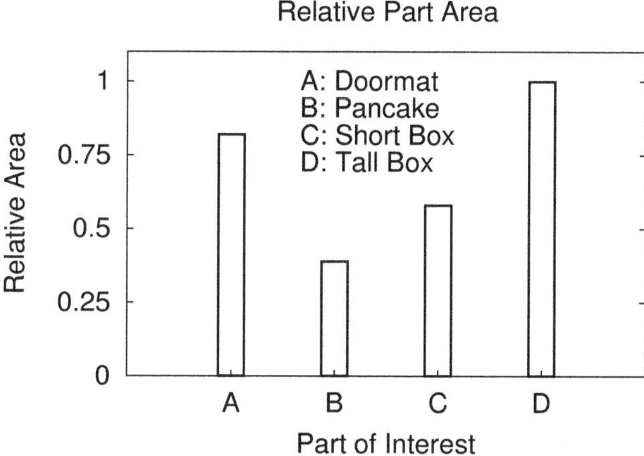

Figure 7.60: This bar graph shows the relative part area for the top part for each of the synthetic objects.

So, this measure might be useful for the parts in Fig. 7.55(c). The problem with this measure, however, is that it does not discriminate the size of a part. Whereas the part area measure is completely reliant on size, the depth measure offers no insight to the part size. Consequently, the pancake, short box, and tall box parts have almost identical values even though they are quite different in size. This measure does not distinguish among these three objects, and yet a human observer would most likely argue that they are different. Human perception would factor their relative size in evaluating their quality as a part. As a practical example, this measure gives very similar results for the screw in Fig. 7.56(b) and the plugs at the top of the distributor cap in Fig. 7.51(b).

The Part Saliency Metric takes a different approach from either of these two measures. Our proposed measure attempts to model human visual perception. Most human observers would rank the top part of the four objects in a manner similar to the graph values in Fig. 7.62. These values are the results from the Part Saliency Metric algorithm. The agreement between this graph and human perception is the power of this algorithm over the area and depth measures. The measure allows one to readily distinguish among the four objects unlike the depth measure, which gave equivalence to the pancake, short box, and tall box parts. The proposed measure factors the size of the part into the estimation of its saliency, yet the measure does not totally consider the size as the part area measure does. The saliency measure also includes protrusion and boundary strength factors such that the doormat receives the smallest saliency value.

These three graphs in Figs. 7.60 through 7.62 motivate our interest and design of the Part Saliency Metric. This measure allows us to quantify and evaluate the quality of each part in a segmentation and thus to merge parts that do not meet a minimum salience requirement. We are able to improve our minima rule decompositions with this measure. We next study the computational performance of the algorithm for this measure.

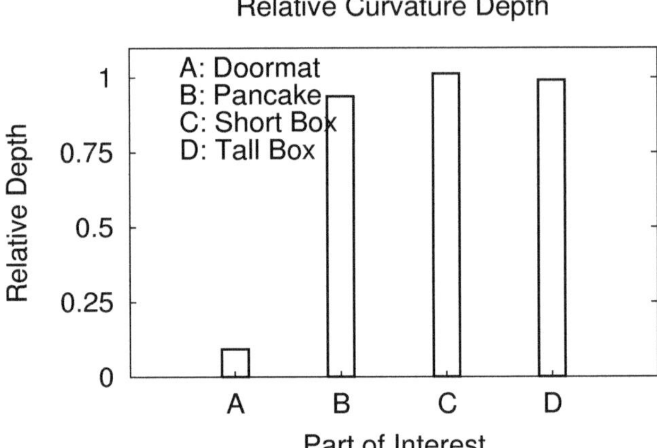

Figure 7.61: This bar graph shows the relative depth in terms of the watershed catchment basin for the top part for each of the synthetic objects. Recall that the segmentation algorithm is a watershed method.

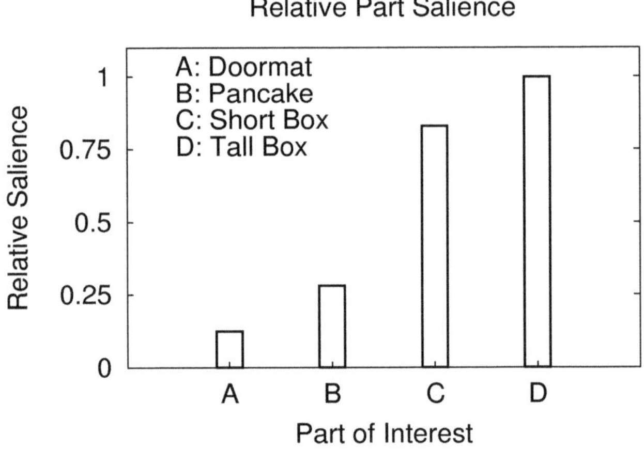

Figure 7.62: This bar graph shows the relative part salience for the top part for each of the synthetic objects.

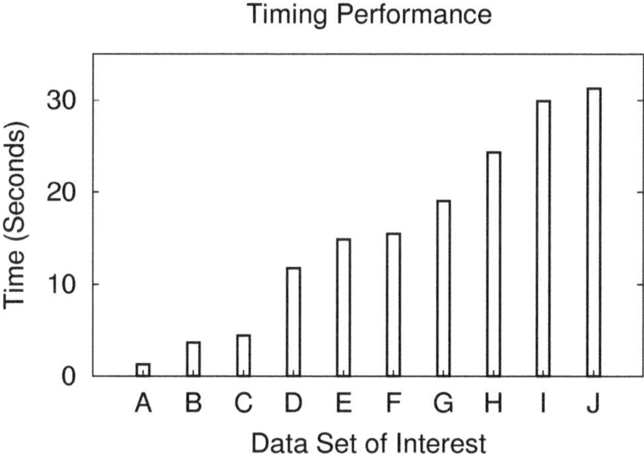

Figure 7.63: This bar graph shows the time in seconds required for the Part Saliency Metric algorithm. The graph shows timing performance for a variety of meshes that are different sizes in terms of triangle count and consist of different numbers of parts.

7.4.3 Timing Performance

We finally explore the timing performance of the proposed algorithm. The bar chart in Fig. 7.63 gives the timing results in seconds for ten different mesh examples. Each bar is the time required to compute the saliency of each part for a particular mesh. The implementation for these results is in Visual C++ on a desktop computer with an Intel Pentium IV at 1.8GHz with 512MB memory. Since the algorithm is a function of the triangle count for a mesh and the number of segmentation parts, we evaluate the performance of the algorithm with a range of examples. Since most objects consist of a small number of parts, the algorithm is almost linear with the number of triangles in the mesh. Although the algorithm has computational complexity of $O(m + n)$ where m is the number of visual parts and n is the number of triangles in the mesh, the algorithm is essentially $O(n)$ since n is often much larger than m. We also note that we have not optimized our implementation. The times reported in the graph could probably be reduced significantly with tuning.

This section concludes our experimental results for the Minima Rule Algorithm and the three algorithms associated with it: Normal Vector Voting, Fast Marching Watersheds, and Part Saliency Metric. This section serves to document the qualitative and quantitative results of each algorithm along with the performance characteristics for each. We now move to the final section of this dissertation, which is the conclusions that we draw from these results and the discussion of future directions for this research.

Table 7.6: Mesh characteristics for timing performance examples.

	Object	Vertices	Triangles	Parts
A	Teapot	$3,034$	$6,010$	5
B	Watering Can	$8,086$	$15,843$	6
C	Coffee Mug	$10,055$	$20,910$	4
D	Chair	$26,766$	$53,462$	11
E	Bore Pin	$37,450$	$74,896$	6
F	Disc Brake	$37,332$	$73,553$	2
G	Hand Crank	$46,870$	$93,752$	7
H	Cone Scene	$61,027$	$117,778$	6
I	Bunny	$34,834$	$69,451$	8
J	Distributor Cap	$65,397$	$129,849$	20

Chapter 8

Conclusions

In this dissertation, we have described an algorithm to decompose triangle meshes that approximate piecewise smooth surfaces into visual parts. This decomposition models the selection of parts that most human observers would choose for the same meshes. This research leverages the human vision theory known as the minima rule. As our final illustration of the algorithm, we once again recall our coffee mug example in Fig. 8.1. In the previous chapters, we have reviewed other research in the literature similar to this algorithm as a context for our contributions, and we have presented the supporting theory along with experimental results to document the algorithm itself. We now conclude this dissertation with a brief summary of the contributions and a short discussion of future directions.

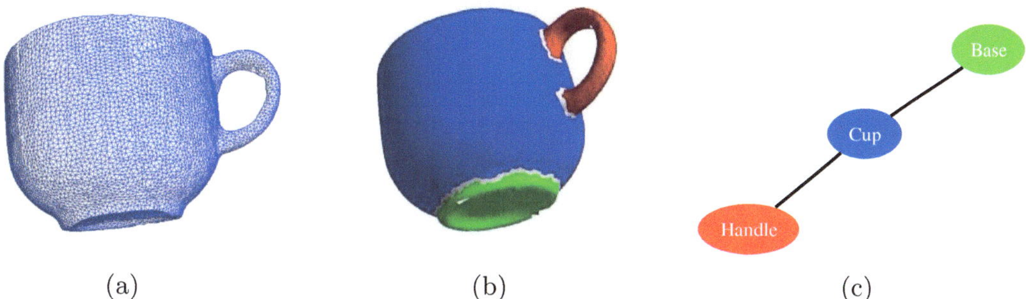

Figure 8.1: This final example shows the decomposition of the mug. (a) The input triangle mesh that approximates the piecewise smooth surface of the coffee mug. (b) The output of the Minima Rule Algorithm that illustrates the visual parts of the mug with color-coded labels. (c) An adjacency graph representation of the decomposition with user-specified labels for the nodes of the graph.

8.1   Summary of Contributions

The primary contribution of this research as described in this dissertation is the creation
of the **Minima Rule Algorithm**. This algorithm extends the state of the art in
computer vision research through the implementation of a human vision theory. Within
this algorithm, we offer three other contributions as follows:

- **Normal Vector Voting**. This contribution is a novel algorithm for the estimation of surface curvature at the vertices of a triangle mesh. The strength of this
  algorithm is its robustness to measurement noise.

- **Fast Marching Watersheds**. This mesh segmentation algorithm identifies negative minima curvature contours to segment a mesh into regions. This contribution follows the definition of visual parts with regard to the minima rule theory.

- **Part Saliency Metric**. This novel measure associates a visual significance, or
  saliency, value with each minima rule part of a mesh. This value allows us to
  merge the least salient parts with other more salient ones and thereby to address
  oversegmentation.

For each of these contributions, we have presented both qualitative and quantitative
results to demonstrate their strengths and to analyze their limitations. We have published the Normal Vector Voting algorithm in (Page et al., 2001; Page et al., 2003e) with
extensions in (Page et al., 2002). We have submitted the Fast Marching Watersheds
to (Page et al., 2003a) and the Part Saliency Metric to (Page et al., 2003b; Page et al.,
2003c) for review. Further, we have submitted a general overview paper to (Page et al.,
2003d). With this summary of the contributions, we now turn to the future directions
for this research.

8.2   Directions for Future Research

The ideas and concepts in this dissertation offer interesting avenues for future research.
Although many directions are possible, we have identified the following areas as particularly important.

8.2.1   Automatic Parameter Selection

The first major area for future research is in automatic selection of the user parameters
associated with each algorithm in this dissertation. Although the algorithms are fairly
insensitive to the parameters in that they do not require fine tuning by the user to
achieve quality results, the requirement for user selection is a minor limitation. Automatic selection of the parameters is not an impossible task, but it does require additional
research. We consider the parameters for each algorithm, individually.

For Normal Vector Voting, we have three parameters as noted in Table 4.2. These
parameters are essentially controls to overcome approximation error in the mesh relative

to the original surface. One solution is statistical analysis of the variation of the mesh. A statistical model could define the appropriate neighborhood size for the voting algorithm and also the decision thresholds for the vertex classification. Another solution is to redesign the voting process. Currently, our scheme is a two step process that requires eigenanalysis at each step. A potential reformulation is to avoid the eigenanalysis during the first step through vector summation instead of the current matrix summation. We have discussed the weaknesses of such an approach in the previous theory sections, but the possibility of overcoming these weaknesses is an interesting avenue for future research. The computational cost for the eigenanalysis is considerable.

The Fast Marching Watersheds algorithm requires two user parameters as noted in Table 5.2. As with the pervious parameters, these variables allow control over the level of robust of the algorithm to errors in curvature estimation. Again, statistical analysis of the data may offer a solution. Another approach is to include additional mathematical morphology operations into the algorithm. For example, one possibility is the skeletonization, or thinning, operation (Rössl et al., 2000). Since we initially threshold the vertex curvatures to establish regions of negative minima curvature, a thinning operation would collapse regions into either a point or a contour. The regions that collapse to a point do not bound a visual part while ones that collapse to a contour do. A simple filtering of point regions would minimize oversegmentation. We need to address a few problems with this approach, however, but it does eliminate the need for the user to select a disc size for the opening and closing operations.

Finally, we consider the parameters for the Part Saliency Metric in Table 6.1. These parameters are more subjective than the previous ones since they are a direct function of human perception. An novel extension to this algorithm would be to avoid these parameters entirely through an artificial neural network solution. The proposed saliency measure is a combination of three measurable quantities of a part: the relative size, the relative protrusion, and the part boundary strength. These values could serve as input to a neural network with the output being the part saliency. The key to this approach is establishing a training set and thus an area for future research.

8.2.2 Object Recognition

Object recognition is an extremely difficult task with most current solutions limited to very constrained and restricted problem domains. Although the Minima Rule Algorithm alone offers no contributions in terms of recognition, as an implementation of a human vision theory, it might serve as a first step in the recognition pipeline. Shapiro and Stockman (Shapiro and Stockman, 2001) suggest commonly used paradigms for object recognition where the method chosen depends heavily on the application. They discuss two paradigms that might be applicable to the results in this dissertation: the matching relational models and the matching functional models. Both of these models use part relationships to move away from a geometric definition of an object to a more symbolic one. Our decomposition algorithm might benefit the creation of such a symbolic representation from a mesh representation.

8.2.3   Complexity Measures

Another avenue for future research is to measure the complexity of an object or scene. Two possible candidate measures from this research are shape complexity and graph complexity.

First, we might use curvature estimates from Normal Vector Voting to evaluate the shape complexity of a surface mesh. Since the algorithm estimates principal curvatures at each vertex, we can generate probability distribution curves. With these curves, we can formulate an information theoretic based on entropy to define shape complexity. In the spirit of Claude Shannon's definition of information, this measure would reflect the amount of shape information that an object possesses. Objects and scenes with nearly constant curvature would contain relatively low values of shape information. While other objects and scenes with significant variation in curvature would exhibit fairly large values. Through some basic experiments, we have already begun to explore these ideas, and currently the selection of appropriate bin widths for estimation of the distributions is a hurdle that we must address. Kernel density estimation is one solution to this issue, but more research is necessary.

A second measure of interest is graph complexity for the Part Adjacency Graph. A common measure of graph complexity is through the matrix-tree theorem (Bondy and Murty, 1976; Chartrand, 1977), which gives the number of nonidentical spanning trees of a graph. This complexity measure gives insight to the number of visual parts and their degree of interconnectedness. Perhaps, we can derive other measures of complexity, but these two are of current interest.

8.2.4   Exponential Map

A side topic that has developed from the geodesic distance work in Normal Vector Voting and Fast Marching Watersheds is the computation of an exponential map for a mesh. With the work of Kimmel and Sethian (Kimmel and Sethian, 1998), we can compute the geodesic distance from one vertex of a mesh to another. A natural extension might be to also compute the departure angle of the geodesic path from one vertex to another. If we can compute this angle, then we can create a geodesic polar mapping (O'Neill, 1997). The problem is that Kimmel and Sethian's method does not preserve angular measures as the algorithm marches across the surface of a mesh. However, the quasi-conformal mappings discussed in (Hurdal et al., 1999) do preserve angular values. The concepts in (Hurdal et al., 1999) along with the straightest geodesic research in (Polthier and Schmies, 1998) may provide a means to compute the departure angle. Subsequently, a combination of these methods with Kimmel and Sethian's work may yield an algorithm to define an exponential map over a triangle mesh.

8.2.5   Discrete Fast Marching Watersheds

The current implementation of the proposed Fast Marching Watersheds algorithm is a continuous implementation. The term continuous means that the values for the height

(a)   (b)

Figure 8.2: The current implementation of the part decomposition algorithm does not necessarily lead to visually pleasing boundaries in terms of smoothness. These zoom views, both (a) and (b) show the jagged nature of the final part boundaries, which are the white triangles in views. A direction of future research is to smooth these boundaries.

map have real values from zero to infinity. The initial motivation for this continuous approach is that our curvature data is also continuous. A potential modification, however, is to assign these continuous height levels—via some constraint—to discrete integer levels. This extension is a discrete implementation of Fast Marching Watersheds.

The advantage is that a discrete definition of the height map suppresses variations in the curvature values and thereby improves the segmentation results. A tradeoff exists, though. Fewer levels lead to greater rejection of curvature noise, but more levels lead to better identification of the negative curvature minima. The success of the method thus requires a balance. Our future research is to investigate this balance.

In addition to noise suppression, the method may also improve the boundaries between parts. Currently, the contours that form the boundaries strictly follow the negative curvature minima, which are often jagged and not visually pleasing. The examples in Fig. 8.2 demonstrate this problem. Although the Fast Marching Watersheds algorithm properly identifies the boundaries, it does so without regard to smoothing constraints. A discrete definition of the height in conjunction with the principles of geodesic erosion (Vincent and Soille, 1991; Sun et al., 2002a) may produce smoother boundary contours while still remaining close to the minima. The distance constraint of geodesic erosion should incorporate some boundary smoothing into the part segmentations.

8.2.6 Visualization Applications

The final topic for future research is the incorporation of the decomposition into mesh simplification algorithms. In 3D computer graphics, these algorithms are very common.

They attempt to reduce the number of triangles in a mesh without loss of fidelity in the visual appearance of the mesh. Our Minima Rule Algorithm may benefit these algorithms. A simple illustration in Fig. 8.3 shows the potential of this approach. Without decomposition, the simplification algorithms require more triangles around the part boundaries to preserve the fidelity of the boundaries. With decomposition, the boundaries do not constrain the simplification process. The application of our decompositions to mesh simplification is an interesting topic of future research.

8.3 Discussion with Closing Remarks

In the first chapter of this dissertation, we began with a mental exercise involving the coffee mug. With this exercise, we suggested that most human observers decompose the mug into three visual parts: the cup, the handle, and the base. Throughout this dissertation, we have explored the minima rule as one potential theory that partially explains our selection of these parts. Further, we have developed the Minima Rule Algorithm in the spirit of this theory as a computer vision algorithm to model human perception. Obviously, the minima rule and our implementation of that rule do not completely capture the perceptual power of the human mind, but hopefully the concepts presented in this dissertation do provide a step—if only a small step—towards extending the state of the art in computer vision.

> "A JOURNEY OF A THOUSAND MILES BEGINS WITH A SINGLE STEP."
> —CONFUCIUS

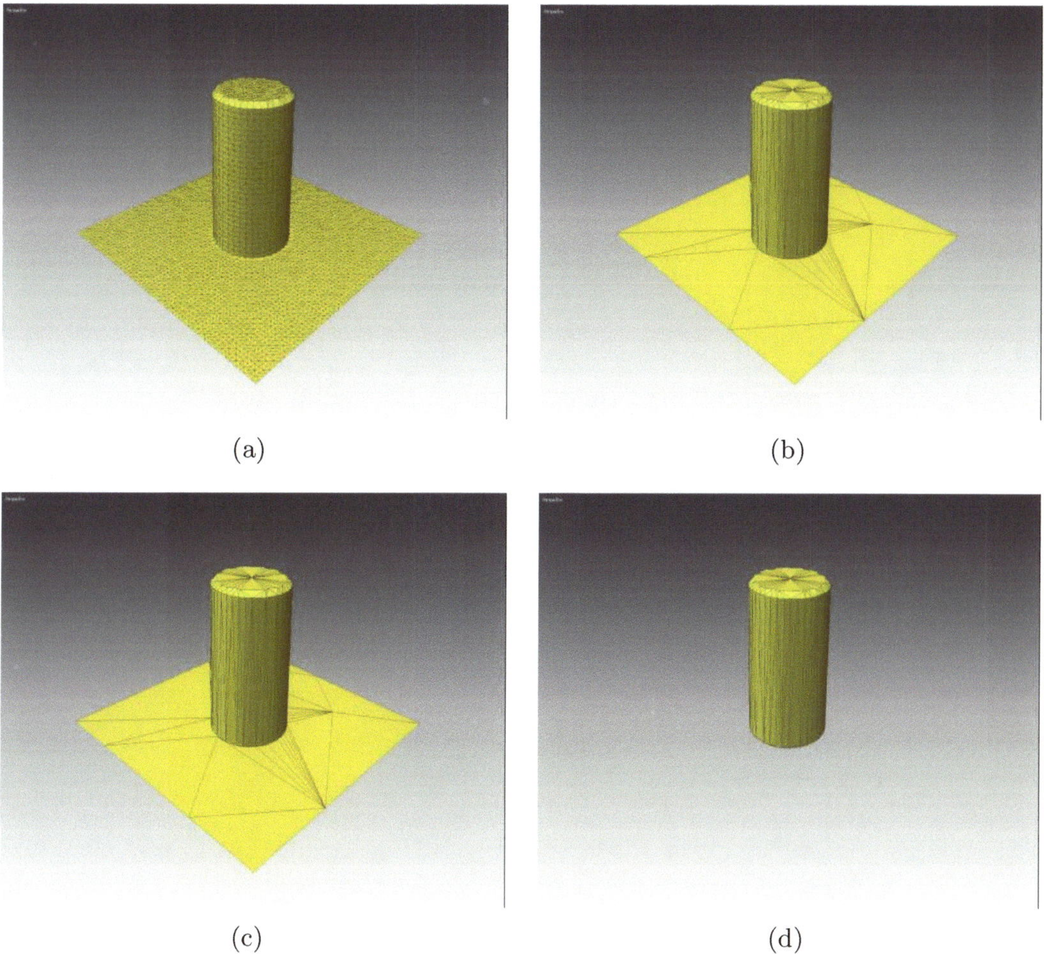

Figure 8.3: Part decomposition such as the Minima Rule Algorithm may aid mesh simplification algorithms. These meshes illustrate the benefit of segmenting the original mesh into a floor and barrel. As a part of the decomposition, we have also applied a hole filling algorithm to fill the boundary holes after decomposition. (a) The original mesh requires over 10K triangles. (b) Simplification without part decomposition. (c) Decomposition into a floor part with simplification. (d) Decomposition into a barrel part with simplification. Both (c) and (d) together require 250 triangles, and (b) also requires 250 triangles. Notice that the floor in (c), however, only requires three triangles while the one in (b) many more than that.

Works Cited

3D Digital Corporation (2000). *RealScan USB White Paper.* `http://www.3ddigitalcorp.com`.

Alrashdan, A., Motavali, S., and Fallahi, B. (2000). Automatic segmentation of digitized data for reverse engineering applications. *IIE Transactions*, 32:59–69.

Amenta, N., Bern, M., and Kamvysselis, M. (1998). A new voronoi-based surface reconstruction algorithm. In *Computer Graphics Proceedings (SIGGRAPH '98)*, pages 415–421.

Andreson, J. R. (1995). *Cognitivie Psychology and its Implications.* W. H. Freeman and Company, New York, NY, fourth edition.

Appel, K. and Haken, W. (1977a). Every planar map is four colorable, part I: Discharging. *Illustrated Journal of Mathematics*, 21.

Appel, K. and Haken, W. (1977b). Every planar map is four colorable, part II: Reducibility. *Illustrated Journal of Mathematics*, 21.

Athitsos, V. and Sclaroff, S. (2002). An appearance-based framework for 3D hand shape classification and camera viewpoint estimation. In *Proceedings of the IEEE Conference on Face and Gesture Recognition*.

Baccar, M., Gee, L. A., Gonzalez, R. C., and Abidi, M. A. (1996). Segmentation of range images via data fusion and morpholocial watersheds. *Pattern Recognition*, 29(10):1671–1685.

Bajaj, C. L. and Dey, T. K. (1992). Convex decomposition of polyhedra and robustness. *SIAM Journal of Computing*, 21(2):339–364.

Bajcsy, R. and Solina, F. (1987). Three-dimensional object representation revisited. In *Proceedings of the First International Conference on Computer Vision*, pages 231–240, London, UK. IEEE Computer Society Press.

Barr, A. H. (1981). Superquadrics and angle-perserving transformations. *IEEE Transactions on Computer Graphics and Applications*, 1(1):11–23.

Baylis, G. C. and Driver, J. (1995a). Obligatory edge assignment in vision–the role figure and part segmentation in symmetry detection. *Journal of Experimental Psychology: Human Perception and Performance*, 21:1323–1342.

Baylis, G. C. and Driver, J. (1995b). One-sided edge assignment in vision. 1. Figure-ground segmentation and attention to objects. *Current Directions in Psychological Science*, 4:140–146.

Bernardini, F., Mittleman, J., Rushmeier, H., Silva, C., and Taubin, G. (1999). The ball-pivoting algorithm for surface reconstruction. *IEEE Transactions on Visualization and Computer Graphics*, pages 349–359.

Besl, P. J. and Jain, R. C. (1988). Segmentation through variable-order surface fitting. *IEEE Transactions on Pattern Analysis and Machine Intelligence*, 10(2):167–192.

Biederman, I. (1985). Human image understanding: Recent research and a theory. *Computer Vision, Graphics and Image Processing*, 32(1):29–73.

Biederman, I. (1987). Recognition by components: A theory of human image understanding. *Psychological Review*, 94(2):115–147.

Biermann, H., Levin, A., and Zorin, D. (2000). Piecewise smooth subdivision surfaces with normal control. In *Computer Graphics Proceedings (SIGGRAPH 2000)*, pages 113–120.

Binford, T. O. (1971). Visual perception by computer. In *IEEE Systems Science and Cybernetics Conference*, Miami, FL.

Blum, H. (1973). Biological shape and visual science. *Journal of Theoretical Biology*, 38:205–287.

Blum, H. and Nagel, R. N. (1978). Shape description using weighted symmetric axis features. *Pattern Recognition*, 10:167–180.

Bondy, J. A. and Murty, U. S. R. (1976). *Graph Theory with Applications*. London, UK.

Braunstein, M. L., Hoffman, D. D., and Saidpour, A. (1989). Parts of visual objects: an experimental test of the minima rule. *Perception*, 18:817–826.

Brinkhoff, T., Kriegel, H.-P., Schneider, R., and Braun, A. (1995). Measuring the complexity of polygonal objects. In *Proceedings of the Third ACM International Workshop on Advances in Geographical Information Systems*, pages 109–117.

Brooks, R. A. (1981). Symbolic reasoning among 3-D models and 2-D images. *Artificial Intelligence*, 17(1–3):285–348.

Bryson, N. and Mobolurin, A. (2000). Towards modeling the query processing relevant shape complexity of 2D polygonal spatial objects. *Information and Software Technology*, 42:357–365.

Burdea, G. and Coiffet, P. (1994). *Vitrual Reality Technology*. John Wiley and Sons, Inc.

Burgiss, S. G., Whitaker, R. T., and Abidi, M. A. (1998). Range image segmentation through pattern analysis of the multiscale wavelet transform. *Digital Signal Processing*, 8:267–276.

Campbell, R. J. and Flynn, P. J. (2001). A survey of free-form object representation and recognition techniques. *Computer Vision and Image Understanding*, 81:166–201.

Castleman, K. R. (1996). *Digital Image Processing*. Prentice–Hall, Inc., Englewood Cliffs, NJ.

Cave, C. B. and Kosslyn, S. M. (1993). The role of parts and spatial relations in object identification. *Perception*, 22(2):229–248.

Cesar, R. M. and Costa, L. (1997). Application and assessment of multiscale bending energy for morphometric characterization of neural cells. *Review of Scientific Instruments*, 68(5):2177–2186.

Chartrand, G. (1977). *Introductory Graph Theory*. Dover Publications, Inc., New York, NY.

Chazelle, B. (1984). Convex partitions of polyhedra: a lower bound and worst-case optimal algorithm. *SIAM Journal of Computing*, 13(3):488–507.

Chazelle, B., Dobkin, D. P., Shouraboura, N., and Tal, A. (1997). Strategies for polyhedral surface decomposition: An experimental study. *Computational Geometry: Theory and Applications*, 7(5–6):327–342.

Chazelle, B. and Incerpi, J. (1984). Triangulation and shape complexity. *ACM Transactions on Graphics*, 3:135–152.

Chazelle, B. and Palios, L. (1990). Triangulating a nonconvex polytope. *Discrete and Computational Geometry*, 5:505–526.

Chazelle, B. and Palios, L. (1997). Decomposing the boundary of a nonconvex polyhedron. *Algorithmica*, 17:245–265.

Chen, X. and Schmitt, F. (1992). Intrinsic surface properties from surface triangulation. In *Proceedings of the European Conference on Computer Vision*, pages 739–743, Santa Margherita Ligure, Italy.

Curless, B. and Levoy, M. (1996). A volumetric method for building complex models from range images. In *Computer Graphics Proceedings (SIGGRAPH '96)*, pages 303–312.

Curless, B. L. (1997). *New Methods for Surface Reconstruction from Range Images*. PhD thesis, Stanford University.

Delingette, H. (1994a). Simplex meshes: A general representation for 3D shape reconstruction. In *Proceedings of the International Conference on Computer Vision and Pattern Recognition*, Seattle, WA.

Delingette, H. (1994b). Simplex meshes: A general representation for 3D shape reconstruction. Technical Report 2214, INRIA.

Delingette, H. (1997). General object reconstruction based on simplex meshes. Technical Report 3111, INRIA.

Delingette, H. (1999). General object reconstruction based on simplex meshes. *International Journal of Computer Vision*, 32(2):111–146.

Desbrun, M., Meyer, M., Schröder, P., and Barr, A. H. (1999). Implicit fairing of irregular meshes using diffusion and curvature flow. In *Computer Graphics Proceedings (SIGGRAPH '99)*, pages 317–324.

Dickinson, S. J., Pentland, A. P., and Rosenfeld, A. (1992). From volumes to views: An approach to 3d object recognition. *Computer Vision, Graphics, and Image Processing: Image Understanding*, 55(2):130–154.

Dijkstra, E. W. (1959). A note on two problems in connexion with graphs. *Numerische Mathematik*, 1:269–271.

do Carmo, M. P. (1976). *Differential Geometry of Curves and Surfaces*. Prentice–Hall, Inc., Englewood Cliffs, NJ.

Driver, J. and Baylis, G. C. (1995). One-sided edge assignment in vision. 2. Part decomposition, shape description, and attention to objects. *Current Directions in Psychological Science*, 4:201–206.

Eck, M. and Hoppe, H. (1996). Automatic reconstruction of B-spline surfaces of arbitrary topological type. In *Computer Graphics Proceedings (SIGGRAPH '96)*, pages 325–334.

Edelsbrunner, H. and Mücke, E. P. (1994). Three-dimensional alpha shapes. *ACM Transactions on Graphics*, 13(1):43–72.

Faber, P. and Fisher, B. (2002). How can we exploit typical architectural structures to improve model recovery? In Cortelazzo, G. M. and Guerra, C., editors, *Proceedings of the International Symposium on 3D Data Processing, Visualization, and Transmission*, pages 824–833, Padova, Italy.

Falcidieno, B. and Spagnuolo, M. (1992). Polyhderal surface decomposition based on curvature analysis. In Kunii, T. L. and Shinagawa, Y., editors, *Modern Geometric Computing for Visualization*, pages 57–72. Springer-Verlag.

Fan, T., Medioni, G., and Nevatia, R. (1986). Description of surfaces from range data using curvature properties. In *Proceedings of the International Conference on Computer Vision and Pattern Recognition*, pages 86–91.

Faugeras, O. D. and Hebert, M. (1986). The representation, recognition and locating of 3-D objects. *International Journal of Robotics Research*, 5(3):27–52.

Ferrie, F. and Levine, M. (1988). Deriving coarse 3D models of objects. In *Proceedings International Conference on Computer Vision and Pattern Recognition*, pages 345–353.

Ferrie, F. P., Lagarde, J., and Whaite, P. (1993). Darboux frames, snakes, and superquadrics: Geometry from the bottom up. *IEEE Transactions on Pattern Analysis and Machine Intelligence*, 15(8):771–784.

Fischler, M. and Bolles, R. (1986). Perceptual organization and curve partitioning. *IEEE Transactions on Pattern Analysis and Machine Intelligence*, 8:100–105.

Flynn, P. J. and Jain, A. K. (1988). Surface classification: Hypothesis testing and parameter estimation. In *Proceedings of the International Conference on Computer Vision and Pattern Recognition*, pages 261–267.

Flynn, P. J. and Jain, A. K. (1989). On reliable curvature estimation. In *Proceedings of the International Conference on Computer Vision and Pattern Recognition*, pages 110–116.

Froimovich, G., Rivlin, E., and Shimshoni, I. (2002). Object classification by functional parts. In Cortelazzo, G. M. and Guerra, C., editors, *Proceedings International Symposium on 3D Data Processing, Visualization, and Transmission*, pages 648–655, Padova, Italy.

Frueh, C. and Zakhor, A. (2001). 3D model generation of cities using aerial photographs and ground level laser scans. In *International Conference on Computer Vision and Pattern Recognition*, volume II, pages 31–38.

Frueh, C. and Zakhor, A. (2002). Data processing algorithms for generating textured 3D building facade meshes from laser scans and camera images. In Cortelazzo, G. M. and Guerra, C., editors, *Proceedings International Symposium on 3D Data Processing, Visualization, and Transmission*, pages 834–847, Padova, Italy.

Gansner, E. R. and North, S. C. (2000). An open graph visualization system and its applications to software engineering. *Software Practice and Experience*, 30(11):1203–1233.

Garland, M. and Heckbert, P. S. (1997). Surface simplification using quadric error metrics. In *Computer Graphics Proceedings (SIGGRAPH '97)*, pages 209–216.

Ghali, A., Daemi, M. F., and Mansour, M. (1998). Image structural information assessment. *Pattern Recognition Letters*, 19:447–457.

Gonzalez, R. C. and Woods, R. E. (1993). *Digital Image Processing*. Addison–Wesley Publishing Company, Reading, MA.

Gonzalez, R. C. and Woods, R. E. (2002). *Digital Image Processing*. Addison–Wesley Publishing Company, Reading, MA, second edition.

Gopi, M., Krishnan, S., and Silva, C. T. (2000). Surface reconstruction based on lower dimensional localized Delaunay triangulations. In Gross, M. and Hopgood, F. R. A., editors, *Computer Graphics Forum (Eurographics 2000)*, volume 19(3), pages C467–478,544.

Gourley, C. S. (1998). *Pattern vector based reduction of large multimodal data sets for fixed rate interactivity during visualization of multiresolution models*. PhD thesis, University of Tennessee, Knoxville, TN.

Gregory, A., State, A., Lin, M. C., Manocha, D., and Livingston, M. A. (1999). Interactive surface decomposition for polyhedral morphing. *Visual Computer*, 15:453–470.

Guzman, A. (1971). Analysis of curved line drawings using context and and global information. In *Machine Intelligence*, volume 6, pages 325–375, Edinburgh, UK. Edinburgh University Press.

Hagen, H., Heinz, S., Thesing, M., and Schreiber, T. (1998). Simulation based modeling. *International Journal of Shape Modeling*, 4(3,4):143–164.

Hameiri, E. and Shimshoni, I. (2002). Using principal curvatures and darboux frame to recover 3D geometric primitives from range images. In *Proceedings International Symposium on 3D Data Processing, Visualization, and Transmission*.

Heawood, P. J. (1890). Map colour theorems. *Quarterly Journal of Mathematics*, 24:332–338.

Hebert, M., Hoffman, R., Johnson, A., and Osborn, J. (1995). Sensor-based interior modeling. In *American Nuclear Society Sixth Topical Meeting on Robotics and Remote Systems*, pages 731–737.

Heckbert, P. S. and Garland, M. (1999). Optimal triangulation and quadric-based surface simplification. *Computational Geometry: Theory and Applications*, 14:49–65.

Heijmans, H. J. A. M. (1994). *Morphological Image Operators*. Academic Press, Boston, MA.

Hershberger, J. E. and Snoeyink, J. S. (1998). Erased arrangements of lines and convex decompositions of polyhedra. *Computational Geometry: Theory and Applications*, 9:129–143.

Hoffman, D. D. (1983). The interpretation of visual illusions. *Scientific American*, 249(6):154–162.

Hoffman, D. D. and Richards, W. A. (1984). Parts of recognition. *Cognition*, 18:65–96.

Hoffman, D. D. and Singh, M. (1997). Salience of visual parts. *Cognition*, 63:29–78.

Hoffman, R. L. and Jain, A. K. (1987). Segmentation and classification of range images. *IEEE Transactions on Pattern Analysis and Machine Intelligence*, 9(5):608–620.

Hoover, A., Goldgof, D., and Bowyer, K. W. (1998). Dynamic-scale model construction from range imagery. *IEEE Transactions on Pattern Analysis and Machine Intelligence*, 20(12):1352–1357.

Hoover, A., Jean-Baptiste, G., and Jiang, X. (1996). An experimental comparison of range image segmentation algorithms. *IEEE Transactions on Pattern Analysis and Machine Intelligence*, 18(7):673–689.

Hoppe, H., DeRose, T., Duchamp, T., Halstead, M., Hubert, J., McDonald, J., Schweitzer, J., and Stuetzle, W. (1994). Piecewise smooth surface reconstruction. In *Computer Graphics Proceedings (SIGGRAPH '94)*, pages 295–302.

Hoppe, H., DeRose, T., Duchamp, T., McDonald, J., and Stuetzle, W. (1992). Surface reconstruction from unorganized points. In *Computer Graphics Proceedings (SIGGRAPH '92)*, volume 26, pages 71–78.

Hughes, C. and Hughes, T. (1999). *Mastering the Standard C++ Classes: An Essential Reference*. John Wiley and Sons, Inc., New York, NY.

Hurdal, M. K., Bowers, P. L., Stephenson, K., Sumners, D. W. L., Rehm, K., Schaper, K., and Rottenberg, D. A. (1999). Quasi-conformally flat mapping the human cerebellum. In Taylor, C. and Colchester, A., editors, *Lecture Notes in Computer Science*, volume 1679, pages 279–286, Berlin. Springer-Verlag. Medical Image Computing and Computer-Assisted Intervention, MICCAI'99.

Integrated Vision Products (2000). *User Documentation: MAPP Ranger System*. Sweden.

Johnson, A., Leger, P., Hoffman, R., Hebert, M., and Osborn, J. (1995). 3-D object modeling and recognition for telerobotic manipulation. In *Proceedings of the Intelligent Robots and Systems*, pages 104–110.

Juttner, M., Caelli, T., and Rentschler, I. (1996). Recognition-by-parts: A computational approach to human learning and generalization of shapes. *Biological Cybernetics*, 74:521–535.

Kanizsa, G. (1979). *Organization in Vision: Essays on Gestalt Perception.* Praeger Publishers, New York, NY.

Karchaher, S., Ritter, D., and Häusler (1997). SLIM3D: software for reverse engineering. Technical report, Friedrich-Alexander-Universität Erlangen. In Lehrstuhl für Optik, `http://www.3d-shape.com/slim/slim_e.html`.

Kettner, L. (1999). Using generic programming for designing a data structure for polyhedral surfaces. *Computational Geometry: Theory and Applications*, 13:65–90.

Kimmel, R. and Sethian, J. A. (1998). Computing geodesic paths on manifolds. In *Proceedings of the National Acadmeny of Sciences*, volume 95, pages 8431–8435.

King, D. and Rossignac, J. (1999). Optimal bit allocation in compressed 3D models. *Computational Geometry: Theory and Applications*, 14:91–118.

Kinsey, L. C. (1993). *Topology of Surfaces.* Springer-Verlag, New York, NY.

Koenderink, J. J. and van Doorn, A. J. (1982). The shape of smooth objects and the way contours end. *Perception*, 11(2):129–137.

Koffka, K. (1935). *Principles of Gestalt Psychology.* Routledge and Kegan Paul, London, UK.

Levoy, M., Rusinkiewicz, S., Ginzton, M., Ginsberg, J., Pulli, K., Koller, D., Anderson, S., Shade, J., Curless, B., Pereira, L., Davis, J., and Fulk, D. (2000). The Digital Michelangelo Project: 3D scanning of large statues. In *Computer Graphics Proceedings (SIGGRAPH 2000)*, pages 131–144.

Li, B. (1993). The moment calculation of polyhedra. *Pattern Recognition*, 26(8):1229–1233.

Li, X., Woon, T. W., Tan, T. S., and Huang, Z. (2001). Decomposing polygon meshes for interactive applications. In *Proceedings of the ACM Symposium on Interactive 3D Graphics*, pages 35–42, North Carolina.

Lin, C. and Perry, M. J. (1982). Shape description using surface triangulation. In *Proceedings of the IEEE Workshop on Computer Vision: Representation and Control*, pages 38–43.

Lindstrom, P. and Turk, G. (1998). Fast and memory efficient polygonal simplification. In *Proceedings Visualization '98*, pages 279–286.

Lingas, A. (1982). The power of non-rectilinear holes. In *Proceedings of the NinethInternational Colloquim on Automata, Languages, and Programming, Lecture Notes in Computer Science*, pages 369–383, New York, NY. Springer-Verlag.

Lorenson, W. and Cline, H. (1987). Marching cubes: a high resolution 3D surface construction algorithm. In *Computer Graphics Proceedings (SIGGRAPH '87)*, pages 163–169.

Mandelbrot, B. (1967). How long is the coast of Britain: statistical self-similarity and fractal dimension. 155:636–638.

Mangan, A. P. and Whitaker, R. T. (1999). Partitioning 3D surface meshes using watershed segmentation. *IEEE Transactions on Visualization and Computer Graphics*, 5(4):308–321.

Marr, D. (1982). *Vision: A Computational Investigation into the Human Representation and Processing of Visual Information*. W. H. Freeman and Company, New York, NY.

Marr, D. and Nishihara, K. (1978). Representation and recognition of the spatial organization three-dimensional shapes. In *Proceedings of the Royal Society of London*, volume 200 of *B*, pages 269–294.

Martin, R. R. (1998). Estimation of principal curvatures from range data. *International Journal of Shape Modeling*, 4(3–4):99–109.

McCallum, B., Nixon, M., Price, B., and Fright, R. (1998). Hand-held laser scanning in practice. In *Image and Vision Computing New Zealand*, pages 17–22. The University of Auckland. `http://www.aranz.com/research/hls/theory.html`.

Medioni, G., Lee, M., and Tang, C. K. (2000). *A Computational Framework for Segmentation and Grouping*. Elsevier, Amsterdam.

Mencl, R. and Müller, H. (1998). Graph-based surface reconstruction using sturctures in scattered point sets. In *Proceedings of the Computer Graphics International '98*, Hanover, Germany.

Mitchell, J. S. B., Mount, D. M., and Papadimitriou, C. H. (1987). The discrete geodesic problem. *SIAM Journal of Computing*, 16(4):647–668.

Mortenson, M. E. (1997). *Geometric Modeling*. John Wiley and Sons, Inc., New York, NY, second edition.

Motavalli, S., Suharitdamrong, V., and Alrashdan, A. (1998). Design model generation for reverse engineering using multi-sensors. *IIE Transactions*, 30:357–366.

Naik, S. M. and Jain, R. C. (1988). Spline-based surface fitting on range images for CAD applications. In *Proceedings of the International Conference on Computer Vision and Pattern Recognition*, pages 249–253.

Neider, J., Davis, T., and Woo, M. (1993). *OpenGL Programming Guide: The official guide to Learning OpenGL, Release 1*. Addison–Wesley Publishing Company, Reading, MA.

Nevatia, R. and Binford, T. O. (1977). Description and recognition of curved objects. *Artificial Intelligence*, 8:77–98.

Oddo, L. A. (1992). Global shape entropy: A mathematically tractable approach to building extraction in aerial imagery. In *Proceedings of the 20th SPIE AIPR Workshop*, volume 1623, pages 91–101.

O'Neill, B. (1997). *Elementrary Differential Geometry*. Academic Press, Orlando, FL, second edition.

O'Rourke, J. (1994). *Computational Geometry in C*. Cambridge University Press.

Osada, R., Funkhouser, T., Chazelle, B., and Dobkin, D. (2002). Shape distributions. *ACM Transactions on Graphics*, 21(4):807–832.

Page, D. L., Koschan, A., and Abidi, M. (2003a). Perception-based 3D triangle mesh segmentation using fast marching watersheds. In *Proceedings of the International Conference on Computer Vision and Pattern Recognition*. To appear.

Page, D. L., Koschan, A., and Abidi, M. (2003b). Progressive perceptual parts: Part decomposition and salience of 3D triangle meshes. In *Computer Graphics Proceedings (SIGGRAPH 2003)*. Submitted for review.

Page, D. L., Koschan, A., Sukumar, S. R., Roui-Abidi, B., and Abidi, M. (2003c). Shape analysis algorithm based on information theory. In *Proceedings of the International Conference on Image Processing*. Submitted for review.

Page, D. L., Koschan, A. F., Sun, Y., Zhang, Y., Paik, J. K., and Abidi, M. A. (2003d). Towards computer-aided reverse engineering of heavy vehicle parts using laser range imaging techniques. *International Journal of Heavy Vehicle Systems*. Submitted to special issue for review.

Page, D. L., Sun, Y., Koschan, A., Paik, J., and Abidi, M. (2001). Robust crease detection and curvature estimation of piecewise smooth surfaces from triangle mesh approximations using normal voting. In *Proceedings of the International Conference on Computer Vision and Pattern Recognition*, volume I, pages 162–167.

Page, D. L., Sun, Y., Koschan, A., Paik, J., and Abidi, M. (2002). Simultaneous mesh simplification and noise smoothing of range images. In *Proceedings of the International Conference on Image Processing*, volume III, pages 821–824.

Page, D. L., Sun, Y., Koschan, A., Paik, J., and Abidi, M. (2003e). Normal vector voting: Crease detection and curvature estimation on large, noisy meshes. *Graphical Models*, 64:1–31.

Pal, N. R. and Pal, S. K. (1989). Entropic thresholding. *Signal Processing*, 16:97–108.

Palmer, S. E. (1977). Hierarchical structure in perceptual representation. *Cognitive Psychology*, 9:441–474.

Pentland, A. P. (1986a). Parts: Structured descriptions of shape. In *Proceedings AAAI-86*, pages 695–701, Philadelphia, PA.

Pentland, A. P. (1986b). Perceptual organization and the representation of natural form. *Artificial Intelligence*, 28(3):293–331.

Pentland, A. P. (1987). Recognition by parts. In *Proceedings of the First International Conference on Computer Vision*, pages 612–620, Washington, DC. IEEE Computer Society Press.

Pentland, A. P. (1989). Part segmentation for object recognition. *Neural Computation*, 1(1):82–91.

Perceptron Incorporated (1993). *LASAR Hardware Manual*. 23855 Research Drive, Farmington Hills, Michigan 38335. http://www.perceptron.com.

Pinkall, U. and Polthier, K. (1993). Computing discrete minimal surfaces and their conjugates. *Experimental Mathematics*, 2(1):15–36.

Polthier, K. and Schmies, M. (1998). Straightest geodesics on polyhedral surfaces. In H. C. Hege, K. P., editor, *Mathematical Visualization*, pages 391–409. Springer-Verlag.

Polthier, K. and Schmies, M. (1999). Geodesic flow on polyhedral surfaces. In *Proceedings of the Joint Eurographics and IEEE TCVG Symposium on Visualization*, pages 179–188.

Press, W. H., Teukolsky, S. A., Vetterling, W. T., and Flannery, B. P. (1992). *Numerical Recipes in C: The Art of Scientific Computing*. Cambridge University Press, New York, NY, second edition.

Pulla, S. (2001). Curvature based segmentation for 3-dimensional meshes. Master's thesis, Arizona State University, Tempe, AZ.

Pulla, S., Razdan, A., and Farin, G. (2002). Improved curvature estimation for watershed segmentation of 3-dimensional meshes. *IEEE Transactions on Visualization and Computer Graphics*. In submission.

Pulli, K., Duchamp, T., Hoppe, H., McDonald, J., Shapiro, L., and Stuetzle, W. (1997). Robust meshes from multiple range maps. In *Proceedings of the International Conference on Recent Advances in 3-D Digital Imaging and Modeling*, pages 205–211.

Reif, U. and Schröder, P. (2000). Curvature smoothness of subdivision surfaces. Technical Report TR-00-03, California Institute of Technology.

Rettmann, M. E., Han, X., and Prince, J. L. (2000). Watersheds on the cortical surface for automated sulcal segmentation. In *IEEE Workshop on Mathematical Methods in Biomedical Image Analysis*, pages 20–27.

Rettmann, M. E., Han, X., and Prince, J. L. (2002). Automated sulcal segmentation using watersheds on the cortical surface. *NeuroImage*, 15(2):329–344.

Riegl Laser Measurement Systems (2000). *Laser Mirror Scanner LMS-Z210: Technical Document and User's Instructions*. http://www.riegl.co.at.

Roerdink, J. B. T. M. (1994). *Shape in Picture. Mathematical Description of Shape in Grey-level Images*, chapter Manifold shape: from differential geometry to mathematical morphology, pages 209–223. NATO ASI Series F 126. Springer-Verlag.

Roerdink, J. B. T. M. (1996). Computer vision and mathematical morphology. *Computing*, pages 131–148. Supplement 11 on Theoretical Foundations of Computer Vision.

Roerdink, J. B. T. M. and Meijster, A. (2001). The watershed transform: definitions, algorithms, and parallelization strategies. *Fundamenta Informaticae*, 41:187–228.

Ronse, C. (1989). Fourier analysis, mathematical morphology, and vision. Technical Report WD54, Philips Research Laboratory Brussels, Brussels, Belgium.

Rosch, E., Mervis, C. B., Gray, W. D., Johnson, D. M., and Boyes-Braem, P. (1976). Basic objects in natural categories. *Cognitive Psychology*, 8:382–439.

Rosin, P. L. (1999). Shape partioning by convexity. In *Proceedings of the British Machine Vision Conference*, pages 633–642.

Rössl, C., Kobbelt, L., and Seidel, H.-P. (2000). Extraction of feature lines on triangulated surfaces using morphological operators. In *Proceedings of the AAAI Symposium on Smart Graphics*, pages 71–75.

Roui-Abidi, B. (1995). *Automatic sensor placement for volumetric object characterization*. PhD thesis, University of Tennessee, Knoxville, TN.

Rubin, E. (1958). Figure and ground. In Beardslee, D. C. and Wertheimer, M., editors, *Readings in Perception*, pages 194–203. Van Nostrand, New York, NY. An abridged translation by Micheal Wertheimer of pages 35–101 of *Visuell wahrgenommene Figuren* (translated by Peter Collet into German from the Danish *Synsoplevede Figurer*, Copenhagen: Gyldendalske, 1915). Copenhagen: Gyldendalske, 1921. Translated and printed by permission of the publishers.

Sacchi, R., Poliakoff, J. F., Thomas, P. D., and Häfele, K.-H. (1999). Curvature estimation for segmentation of triangulated surfaces. In *Proceedings of the International Conference on Recent Advances in 3-D Digital Imaging and Modeling*, pages 536–543.

Sander, P. T. and Zucker, S. W. (1986). Stable surface estimation. In *Proceedings of the Eighth International Conference on Pattern Recognition*, pages 1165–1167, Paris, France.

Sander, P. T. and Zucker, S. W. (1990). Inferring surface trace and differential structure from 3-D images. *IEEE Transactions on Pattern Analysis and Machine Intelligence*, 12(9):833–854.

Sanniti di Baja, G. and Svensson, S. (2002). A new shape descriptor for sufaces in 3D images. *Pattern Recognition Letters*, 23:703–711.

Sapidis, N. S. and Besl, P. J. (1995). Direct construction of polynomial surfaces from dense range images through region growing. *ACM Transactions on Graphics*, 14(2):171–200.

Sebastian, R., Clark, R., Simonson, D., and Slotwinski, A. (1995). *Fiber Optic Coherent Laser Radar 3D Vision System.* Coleman Research Corporation, Springfield, VA.

Shannon, C. E. (1948). A mathematical theory of communication. *The Bell System Technical Journal*, 27:379–423, 623–656.

Shapiro, L. G. and Stockman, G. C. (2001). *Computer Vision.* Prentice–Hall, Inc., Upper Saddle River, NJ.

Siddiqi, K. and Kimia, B. B. (1995). Parts of visual form: Computational aspects. *IEEE Transactions on Pattern Analysis and Machine Intelligence*, 17(3):239–251.

Siek, J. G., Lee, L.-Q., and Lumsdaine, A. (1999). The generic graph component library. In *Proceedings of the Conference on Object Oriented Programming Systems, Languages, and Applications*, pages 399–414. ACM Special Interest Group on Programming Languages. `http://www.boost.org/libs/graph/doc/`.

Siek, J. G., Lee, L.-Q., and Lumsdaine, A. (2001). *The Boost Graph Library: User Guide and Reference Manual.* Addison–Wesley Publishing Company.

Singh, M., Seyranian, G. D., and Hoffman, D. D. (1999). Parsing silhouettes: the short-cut rule. *Perception and Psychophysics*, 61(4):636–660.

Stoddart, A. J., Lemke, A., Hilton, A., and Renn, T. (1998). Estimating pose uncertainty for surface registration. *Image and Vision Computing*, 16:111–120.

Stroustrup, B. (1991). *The C++ Programing Language.* Addison–Wesley Publishing Company, Reading, MA, second edition.

Stuwe, M. (1989). *Plateau's Problem and the Calculus of Variations.* Princeton University Press, Princeton, NJ.

Suk, M. and Bhandarkar, S. M. (1992). *Three-Dimensional Object Recognition from Range Images.* Springer-Verlag, Tokyo.

Sun, Y. and Abidi, M. A. (2001). Surface matching by 3D point's fingerprint. In *Proceedings of the Eighth International Conference on Computer Vision*, volume 2, pages 263–269.

Sun, Y., Page, D. L., Koschan, A., Paik, J., and Abidi, M. (2002a). Triangle mesh-based edge detection and its application to surface segmentation and adaptive surface smoothing. In *Proceedings of the International Conference on Image Processing*, volume III, pages 825–828.

Sun, Y., Page, D. L., Paik, J., Koschan, A., and Abidi, M. (2002b). Triangle mesh-based surface modeling using adaptive smoothing and implicit texture integration. In *Proceedings of the International Symposium on 3D Data Processing, Visualization, and Transmission*, pages 588–597.

Svensson, S. and Sanniti di Baja, G. (2001). A tool for decomposing 3D discrete objects. In *International Conference on Computer Vision and Pattern Recognition*, volume I, pages 850–855.

Svensson, S. and Sanniti di Baja, G. (2002). Using distance transforms to decompose 3D discrete objects. *Image and Vision Computing*, 20:529–540.

Tan, T. S., Chong, C. K., and Low, K. L. (1999). Computing bounding volume hierarchy using simpilfied models. In *Proceedings of the ACM Symposium on Interactive 3D Graphics*, pages 63–69.

Tang, C. K. and Medioni, G. (1999). Robust estimation of curvature information from noisy 3D data for shape description. In *Proceedings of the Seventh International Conference on Computer Vision*, pages 426–433, Kerkyra, Greece.

Tang, C. K. and Medioni, G. (2002). Curvature augmented tensor voting for shape inference from noisy 3D data. *IEEE Transactions on Pattern Analysis and Machine Intelligence*, 24(6):858–864.

Tang, K., Chou, S.-Y., Chen, L.-L., and Woo, T. C. (2000). Tetrahedral mesh generation for solids based on alternating sum of volumes. *Computers in Industry*, 41:65–81.

Taubin, G. (1995). Estimating the tensor of curvature of a surface from a polyhedral approximation. In *Proceedings of the Fifth International Conference on Computer Vision*, pages 902–907.

Terzopoulos, D., Witkin, A., and Kass, M. (1987). Symmetry-seeking models and 3D object reconstruction. *International Journal of Computer Vision*, 1(3):211–221.

Thompson, W. B., Owen, J. C., and de St. Germain, H. J. (1999). Feature-based reverse engineering of mechanical parts. *IEEE Transactions on Robotics and Automation*, 15:57–66.

Tookey, R. M. and Ball, A. A. (1997). *The Mathematics of Surfaces VII*, chapter Estimation of curvatures from planar point data, pages 131–144. Information Geometers, Winchester.

Toussaint, G. (1991). Efficient triangulation of simple polygons. *Visual Computer*, 7:280–295.

Trucco, E. and Verri, A. (1998). *Introductory Techniques for 3-D Computer Vision*. Prentice–Hall, Inc., Upper Saddle River, NJ.

Tversky, B. and Hemenway, K. (1984). Objects, parts, and categories. *Journal of Experimental Psychology: General*, 113:169–193.

Vaina, L. M. and Zlateva, S. D. (1990). The largest convex patches: a boundary-based method for obtaining object parts. *Biological Cybernetics*, 62(3):225–236.

van Vliet, L. J. and Verbeeck, P. W. (1993). Curvature and bending energy in digitized 2D and 3D images. In *Proceedings of the Eighth Scandinavian Conference on Image Analysis*, volume 2, pages 1403–1410.

Várady, T. and Hermann, T. (1996). *The Mathematics of Surfaces VI*, chapter Best Fit Surface Curvature at Vertices of Topologically Irregular Curve Networks, pages 411–427. Oxford University Press, Oxford.

Veltkamp, R. (1999). Generic geometric programming in the computational geometry algorithms library. *Computer Graphics Forum*, 18(2). `http://www.cgal.org`.

Vemuri, B., Mitiche, A., and Aggarwal, J. (1986). Curvature-based representation of objects from range data. *Image and Vision Computing*, 4(2):107–114.

Viceconti, M., Casali, M., Massari, B., Cristofolini, L., Bassini, S., and Toni, A. (1996). The standardized femur program: Proposal for a reference geometry to be used for the creation of finite element models of the femur. *Journal of Biomechanics*, 29(1241). `http://www.cineca.it/hosted/LTM-IOR/back2net/ISB_mesh/mesh_list.html`.

Vincent, L. and Soille, P. (1991). Watersheds in digital spaces: an efficient algorithm based on immersion simulations. *IEEE Transactions on Pattern Analysis and Machine Intelligence*, 13(6):583–598.

Weiss, M. A. (1999). *Data Structures and Algorithm Analysis in C++*. Addison–Wesley Publishing Company, Reading, MA, second edition.

Werghi, N. and Xiao, Y. (2002). Posture recognition and segmentation from 3D human body scans. In Cortelazzo, G. M. and Guerra, C., editors, *Proceedings International Symposium on 3D Data Processing, Visualization, and Transmission*, pages 636–639, Padova, Italy.

Wertheimer, M. (1958). Principles in perceptual organization. In Beardslee, D. C. and Wertheimer, M., editors, *Readings in Perception*, pages 115–135. Van Nostrand, New York, NY. An abridged translation by Micheal Wertheimer of *Utersuchungen zur Lehre von der Gestalt*, II. *Psychol. Forsch.*, 1923, 4, 301–350. Translated and printed by permission of the publisher, Springer, Berlin.

Whitaker, R. T. (1996). A level-set approach to 3D reconstruction from range data. Technical Report EE-96-07-01, University of Tennessee, Knoxville, TN. To appear in International Journal of Computer Vision.

White, D., Scribner, K., and Olafsen, E.

Wu, K. and Levine, M. D. (1997). 3D part segmentation using simulated electrical charge distributions. *IEEE Transactions on Pattern Analysis and Machine Intelligence*, 19(11):1223–1235.

Yan, X. and Gu, P. (1996). A review of rapid prototyping technologies and systems. *Computer-Aided Design*, 28(4):307–318.

Yang, M. and Lee, E. (1999). Segmentation of measured point data using a parametric quadric surface approximation. *Computer-Aided Design*, 31:449–457.

Young, I. T., Walker, J. E., and Bowie, J. E. (1974). An analysis technique for biological shape I. *Info Control*, 25:357–370.

Yu, Y., Ferencz, A., and Malik, J. (2001). Extracting objects from range and radiance images. *IEEE Transactions on Visualization and Computer Graphics*, 7(4):351–364.

About the Author

David Lon Page, Ph.D., is a research scientist with a doctorate in electrical engineering from the University of Tennessee. His work centers on computer vision, three-dimensional imaging, and applied scientific research, with contributions spanning peer-reviewed journals, conference proceedings, and collaborative research programs.

He grew up in Tennessee and developed an early interest in both technology and public life. Long before graduate school, he found himself drawn to two parallel pursuits: understanding how systems work and observing how people organize themselves within them. One path led to engineering, mathematics, and scientific inquiry. The other led to campaigns, civic conversations, and a sustained curiosity about American institutions.

Those interests continued to develop through his academic career. Graduate study refined his technical focus while reinforcing a broader habit of inquiry—how complex systems behave, how signals emerge from noise, and how structure shapes outcomes. That perspective, grounded in engineering but extending beyond it, continues to inform his thinking across disciplines.

Outside the laboratory, he has remained an engaged observer of civic life for decades. His interest in politics does not arise from partisanship, but from structure—how institutions channel conflict, how incentives shape behavior, and how durable systems manage disagreement without collapse.

He lives in Knoxville, Tennessee, with his wife, Lisa. Their daughter, Grace, now away at college, has left behind a quieter household and a different rhythm of daily life. The transition has created space for reflection, writing, and the continuation of questions that have followed him for years—questions about systems, society, and the conditions under which both endure.

Selected Technical Publications

Portions of the author's research career have appeared in the following peer-reviewed venues in computer vision and 3D imaging:

- D. Page, Y. Sun, A. Koschan, J. Paik, M. Abidi, *Normal vector voting: crease detection and curvature estimation on large, noisy meshes, Graphical Models*, Vol. 64, pp. 1–31 (2003).

- D. Page, Y. Sun, A. Koschan, J. Paik, M. Abidi, *Simultaneous mesh simplification and noise smoothing of range images*, in *Proceedings of the International Conference on Image Processing*, Vol. III, pp. 821–824 (2002).

- D. Page, Y. Sun, A. Koschan, J. Paik, M. Abidi, *Robust crease detection and curvature estimation of piecewise smooth surfaces from triangle mesh approximations using normal voting*, in *Proceedings of the Conference on Computer Vision and Pattern Recognition*, Vol. I, pp. 162–167 (2001).